CAMBRIDGE LIBRARY COLLECTION

Books of enduring scholarly value

Technology

The focus of this series is engineering, broadly construed. It covers technological innovation from a range of periods and cultures, but centres on the technological achievements of the industrial era in the West, particularly in the nineteenth century, as understood by their contemporaries. Infrastructure is one major focus, covering the building of railways and canals, bridges and tunnels, land drainage, the laying of submarine cables, and the construction of docks and lighthouses. Other key topics include developments in industrial and manufacturing fields such as mining technology, the production of iron and steel, the use of steam power, and chemical processes such as photography and textile dyes.

Life of Robert Stevenson

Published in 1878, this biography of the civil engineer Robert Stevenson (1772–1850) was written by his second-youngest son David (1815–86), also a civil engineer and uncle to the author Robert Louis Stevenson. Having already published *The Principles and Practice of Canal and River Engineering* in 1872 (also reissued in this series), he set about writing this survey of his father's life and works, based on extracts from Robert's professional reports, notes from his diary, and communications to scientific journals and societies between 1798 and 1843. Perhaps most widely known for his practical and persuasive leadership in building many lighthouses for the Northern Lighthouse Board – including that on the notorious Bell Rock, over which he came into conflict with engineer John Rennie regarding the design – Stevenson ensured that the Scottish coastline became a much safer place for shipping for decades to come.

Cambridge University Press has long been a pioneer in the reissuing of out-of-print titles from its own backlist, producing digital reprints of books that are still sought after by scholars and students but could not be reprinted economically using traditional technology. The Cambridge Library Collection extends this activity to a wider range of books which are still of importance to researchers and professionals, either for the source material they contain, or as landmarks in the history of their academic discipline.

Drawing from the world-renowned collections in the Cambridge University Library and other partner libraries, and guided by the advice of experts in each subject area, Cambridge University Press is using state-of-the-art scanning machines in its own Printing House to capture the content of each book selected for inclusion. The files are processed to give a consistently clear, crisp image, and the books finished to the high quality standard for which the Press is recognised around the world. The latest print-on-demand technology ensures that the books will remain available indefinitely, and that orders for single or multiple copies can quickly be supplied.

The Cambridge Library Collection brings back to life books of enduring scholarly value (including out-of-copyright works originally issued by other publishers) across a wide range of disciplines in the humanities and social sciences and in science and technology.

Life of Robert Stevenson

Civil Engineer

DAVID STEVENSON

CAMBRIDGE
UNIVERSITY PRESS

CAMBRIDGE
UNIVERSITY PRESS

University Printing House, Cambridge, CB2 8BS, United Kingdom

Cambridge University Press is part of the University of Cambridge.

It furthers the University's mission by disseminating knowledge in the pursuit of
education, learning and research at the highest international levels of excellence.

www.cambridge.org
Information on this title: www.cambridge.org/9781108070584

© in this compilation Cambridge University Press 2014

This edition first published 1878
This digitally printed version 2014

ISBN 978-1-108-07058-4 Paperback

LIFE OF ROBERT STEVENSON.

Edinburgh University Press:
THOMAS AND ARCHIBALD CONSTABLE, PRINTERS TO HER MAJESTY.

ROBERT STEVENSON F.R.S.E.

CIVIL ENGINEER.

From a bust by Joseph, placed in the Library of the Bell Rock Light house by the Commissioners of the Northern Light houses.

LIFE

OF

ROBERT STEVENSON

CIVIL ENGINEER

FELLOW OF THE ROYAL SOCIETY OF EDINBURGH ; FELLOW OF THE GEOLOGICAL SOCIETY OF LONDON ;
FELLOW OF THE ASTRONOMICAL SOCIETY OF LONDON ; MEMBER OF THE SOCIETY
OF SCOTTISH ANTIQUARIES, OF THE WERNERIAN NATURAL HISTORY
SOCIETY, AND OF THE INSTITUTION OF CIVIL ENGINEERS.

ENGINEER TO THE COMMISSIONERS OF NORTHERN LIGHTHOUSES AND TO
THE CONVENTION OF ROYAL BURGHS OF SCOTLAND, ETC.

BY

DAVID STEVENSON

CIVIL ENGINEER

VICE-PRESIDENT OF THE ROYAL SOCIETY OF EDINBURGH ;
MEMBER OF COUNCIL OF THE INSTITUTION OF CIVIL ENGINEERS, ETC.

ADAM AND CHARLES BLACK, EDINBURGH
E. AND F. N. SPON, LONDON AND NEW YORK
1878.

PREFACE.

THE addresses made to the Royal Society of Edinburgh, and the Institution of Civil Engineers, at the opening meetings of the session—1851, contained obituary notices of Robert Stevenson. The late Alan Stevenson, his eldest son, also wrote a short Memoir of his father, which was printed for private circulation.

But Robert Stevenson's long practice as a Civil Engineer—the important works he executed—and the valuable contributions he made to Engineering and Scientific literature, seem to me to require a fuller notice of his life than has hitherto been given.

This has been attempted in the following Memoir, which will be found to consist of extracts from Mr. Stevenson's Professional Reports—of notes from his Diary—and of communications to Scientific Journals and Societies, between the years 1798 and 1843, when he retired from active practice.

These papers embrace a wide field of Engineering, including Lighthouses, Harbours, Rivers, Roads, Railways, Ferries, Bridges, and other cognate subjects.

Some of them describe Engineering practice which is now obsolete, but not on that account, I think, uninteresting to such modern Engineers as have regard for the antiquities of their Profession.

Some of them, I am aware, can only be appreciated by those who are specially interested in the city of Edinburgh.

All of them will, I venture to think, be found worthy of preservation as interesting Engineering records of an era that has passed away. It formed no part of my duty to criticise them, in the light of modern Engineering, and, unaltered in form of expression or statement of opinion, they are now reproduced as they came from my father's pen.

I offer no apology for presenting these Extracts as the outlines of the life of one who occupied a prominent place among the Civil Engineers who practised during the beginning of the present, and end of the last century, shortly after British Engineering, with Smeaton as its founder, may be said to have had its origin.

D. S.

EDINBURGH, *July* 1878.

CONTENTS.

CHAPTER I.

EARLY LIFE.

CHAPTER II.

BELL ROCK LIGHTHOUSE.

CHAPTER III.

LIGHTHOUSE ILLUMINATION.

CONTENTS.

CHAPTER XVI.

CONTRIBUTIONS ON ENGINEERING AND SCIENTIFIC SUBJECTS.

CHAPTER XVII.

EXTRACTS FROM EARLY REPORTS.

CHAPTER XVIII.

LIST OF PLATES.

CHAPTER I.

EARLY LIFE.

1772—1798.

Birth—Mr. Smith's improvements in Lighthouse illumination—Origin of the Scottish Lighthouse Board—Acts as Assistant to their Engineer—Student at Andersonian Institution, Glasgow, and University of Edinburgh—Succeeds Mr. Smith as Engineer to the Northern Lighthouse Board—Tour of inspection of English lights in 1801—Is taken for a French spy.

ROBERT STEVENSON, maltster in Glasgow, was born in 1720, and, as stated on his tombstone, in the burial-ground of the Cathedral, died in 1764.

His fourth son, Alan, was partner in a West India house in Glasgow, and died of fever in the island of St. Christopher, in 1774, while on a visit to his brother, who managed the foreign business of the house at that place.

The only son of Alan Stevenson was Robert, the subject of this Memoir, who was born at Glasgow on the 8th of June 1772.

When his father died, Robert Stevenson, then an infant, was left in circumstances of difficulty, for the same epidemic fever which deprived him of his father carried off his uncle also, at a time when their loss operated most disadvantageously on the business which they conducted ; and, strange to say, on account of

A

legal difficulties, nearly half a century elapsed before any patrimonial funds in which my father had an interest were realised.

Under these circumstances his mother (Jean Lillie, daughter of David Lillie, builder in Glasgow, who died, as stated on his tombstone, in the Cathedral burying-ground, in 1774) resolved to go to Edinburgh to reside with a married sister, and when her son reached the age of being able for school she wisely took advantage of one of the hospitals in that city for his education; and the spirit of the man is well brought out by the fact that he devoted his first earnings in life, at the Cumbrae Lighthouse, to making a *contribution* to the funds of the Orphan Hospital in payment of what he regarded as a *debt*.

It appears from "Memoranda" left by my father for the information of his family, that his mother was a woman of great prudence and remarkable fortitude, based on deep convictions of religion; and, even in their time of trial, which lasted over his school days, he says, —"My mother's ingenuous and gentle spirit amidst all her difficulties never failed her. She still relied on the providence of God, though sometimes, in the recollection of her father's house and her younger days, she remarked that the ways of Providence were often dark to us. The Bible, and attendance on the ministrations, chiefly of Mr. Randall of Lady Yester's Church, afterwards Dr. Davidson of the Tolbooth,[1] and at other churches,

---

[1] Mr. Randall assumed the name of Davidson after succeeding to the estate of Muirhouse.

where I was almost always her constant attendant, were the great sources of her comfort.

"Her intention was that I should be trained for the ministry, with a view to which I had been sent, after leaving my first school, to Mr. Macintyre, a famous linguist of his day, where I made the acquaintance of Patrick Neill, afterwards the well-known printer, and still better known naturalist, who remained my most intimate friend through life, and of William Blackwood, the no less celebrated publisher."

Circumstances, however, occurred which entirely changed my father's prospects and pursuits. Soon after he had attained his fifteenth year his mother was married to Mr. Thomas Smith—son of a shipowner, and member of the Trinity House of Dundee,—who himself was, my father says, a "furnishing iron-merchant, shipowner, and underwriter" in Edinburgh, and who being also a lamp-maker and an ingenious mechanician, appears at a very early date to have directed his attention to the subject of lighthouses, and endeavoured to improve the mode of illumination then in use, by substituting lamps with mirrors, for the open coal-fires which were at that early time the only beacons to guide the mariner.

Mr. Smith's improvements attracted the notice of Professor Robison, Sir David Hunter Blair, and Mr. Creech, the publisher and honorary secretary to the Chamber of Commerce. I find from the minutes of that body, that in 1786, a complaint was made to them by shipmasters as to the defective state of the coal light on

the Isle of May, which was a "private light" belonging
to the family of the Duke of Portland.

The Chamber sent a deputation of their number to
inquire into the truth of the objections that had been
made, who fully confirmed the justice of the complaints.

When the result of the examination was reported
to the Chamber of Commerce, Mr. Smith submitted to
them "a plan for improving the light by dispensing
with the coal-fire," and after fully considering his suggestions, the Chamber, at their meeting of 24th May 1786,
resolved "that while they allowed much ingenuity to
Mr. Smith's plan of reflectors, they were of opinion that
a coal light should be continued."

The Board of Northern Lighthouses was constituted
by Act of Parliament in 1786; its members were the
Lord Advocate and Solicitor-General, the chief magistrates of Edinburgh, Glasgow, Aberdeen, Inverness, and
Campbeltown, and the Sheriffs of the maritime counties
of Scotland. These Commissioners, happily for the
interests of navigation, took a more enlightened view
of their duties than the Chamber of Commerce of Edinburgh, and after hearing and considering Mr. Smith's
proposals, formally appointed him their Engineer.

The preamble of the Act constituting the Northern
Lighthouse Board, states that it would greatly conduce to
the security of navigation and the fisheries if *four* light-
houses were erected in the north part of Great Britain.
Such, it would seem, was the limited state of trade in
Scotland, that the erection of these four lighthouses was

all that was contemplated, on a coast, extending to about 2000 miles, of perhaps the most dangerous navigation in Europe. It is now marked by sixty lighthouse stations for the guidance of the sailor, but new claims continue to be made, and new lighthouses are still admitted to be required.

The newly established Lighthouse Board at once entered on its important duties, and the first light they exhibited was Kinnaird Head, which was designed by Mr. Smith and lighted in 1787.

These pursuits being very congenial to my father's mechanical turn of mind, he had rendered himself useful to Mr. Smith in carrying them out, and was intrusted, at the early age of nineteen, to superintend the erection of a lighthouse on the island of Little Cumbrae, in the river Clyde, according to a design which Mr. Smith had furnished to the Cumbrae Light Trustees. This connection soon led to his adoption as Mr. Smith's partner in business, and, in 1799, to his union with his eldest daughter by a former marriage.

During the cessation of the works at Cumbrae in winter, my father, who had determined to follow the profession of a Civil Engineer, applied himself, as appears from class note-books in my possession, with great zeal to the practice of surveying and architectural drawing, and to the study of mathematics at the Andersonian Institution at Glasgow. Of the kindness of Dr. Anderson, who presided over that Institution, he ever entertained a most grateful remembrance, and often spoke of him as one of his best advisers and kindest friends, and in the Memoranda already noticed he records his obligations to him in the following words :— "It was the practice of Professor Anderson kindly to befriend and forward the views of his pupils; and his attention to me during the few years I had the pleasure of being known to him was of a very marked kind, for he directed my attention to various pursuits, with the view to my coming forward as an engineer."

After completing the Cumbrae Lighthouse he was further engaged, under Mr. Smith, in erecting two lighthouses on the Pentland Skerries in Orkney, where, in view of what lay before him at the Bell Rock, he had the useful experience of living four months in a tent on an uninhabited island, and arranging the landing of the whole of the materials of the lighthouses in the difficult navigation of the Pentland Firth. But here also he had a personal experience of God's overruling Providence, which clung to him through life, and, as we shall find, proved his stay in times of danger, when personal resources had ceased to prove availing. In returning from the Pentland Skerries, in 1794, he embarked in the sloop 'Elizabeth' of Stromness, and proceeded as far as Kinnaird Head, when the vessel was becalmed about three miles from the shore. The captain kindly landed my father, who continued his journey to Edinburgh by land. A very different fate, however, awaited his unfortunate shipmates. A violent gale came on, which drove the 'Elizabeth' back to Orkney, where she was totally wrecked, and all on board unhappily perished.

Notwithstanding my father's active duties in summer, he was so zealous in the pursuit of knowledge that he contrived, during several successive winters, on his return from his practical work, to avail himself of the Philosophical classes at the University of Edinburgh. In this manner he attended Professor Playfair's second and third Mathematical courses, two sessions of Robison's Natural Philosophy, two courses of Chemistry under Dr. Hope, and two of Natural History under Professor Jameson. To these he added a course of Moral Philosophy under Dugald Stewart, a course of Logic under Dr. Ritchie, and one of Agriculture under Professor Low. "I was prevented, however," he remarks, in the Memoranda, "from following my friend Dr. Neill for my degree of M.A. by my slender knowledge of Latin, in which my highest book was the Orations of Cicero, and by my total want of Greek." Such zeal in the pursuit of knowledge, and views so enlarged of the benefits and value of a liberal education, were characteristics of a mind of no ordinary vigour; so that, early trained to practical work, and inspired with a true love of his profession, it was not unnatural that on the resignation of Mr. Smith the Board should have appointed Mr. Stevenson to succeed him as their Engineer.

The first annual report made by him to the Board is dated June 1798, and he continued annually to prepare one up to the time of his resignation in 1843.

The first occasion on which he was sent by the Board on a special mission was in 1801, when he was deputed by the Commissioners to visit and report on the Lighthouses on the coasts of England, Wales, and the Isle of Man. The report he submitted to the Board is a most elaborate and valuable document. After describing upwards of twenty Public, Private, and Harbour lights which he had examined, he proceeds fully to discuss the different systems of management in use, and particularly to compare the system adopted by the Scotch Board with that practised in England by the Trinity House, most readily advising the adoption of what seemed improvements in the administration of the Southern Board. In reporting as to the Isle of Man he takes occasion to suggest that the lighting of that island should be taken up by the Northern Commissioners—a proposal which was acted on in 1815. He says :—

"I had several communications with William Scott, Esq., Receiver-General of the Customs, upon the subject of Lighthouses. At his request I went to the Point of Langness, and to the Calf of Man; the former a very dangerous point of land, the latter a situation that seems every way answerable to the general purposes of a site for a lighthouse.

"As this island occupies a middle situation between Great Britain and Ireland, and is not included in any of these Acts of Parliament which relate to the erecting or maintaining of Lights, on either side of the Channel, perhaps it might answer to include the Isle of Man under the same Act which refers to the Northern Lighthouses; and by extending your powers this island might no longer stand a monument of darkness, and a great obstruction to the navigation of St. George's Channel, particularly from the want of a light upon the Calf of Man.

"Such a light, together with the late improvement of the

Copeland light, and the erection of the Kilwarlin light upon the Irish coast, would in an eminent degree improve the navigation of the Irish Channel. From the central situation of the Isle of Man, a light would soon pay itself, by serving the trade of Maryport, Workington, Whitehaven, Lancaster and Liverpool, on the one side of the Channel, with Dublin and Newry on the other."

With reference to this suggestion the Commissioners, in January 1802, adopted the following resolution :—

"In the above report Mr. Stevenson has stated very strongly the great utility of a lighthouse upon the Calf of Man; but not being within the jurisdiction either of the Trinity House of London, or of the Commissioners for the Northern Lighthouses, both of them are thereby prevented from accomplishing an object so much wished for by mariners, as it would prove a great additional security to the navigation between a great number of the ports on the west of England, and Dublin, and other ports in Ireland. In order therefore that this circumstance may not be overlooked, the Commissioners directed this notice to be taken of it in their Minutes, in order that if any application to Parliament shall at a future period be deemed necessary, the Commissioners may judge how far it may not be proper to apply for power and liberty to erect a lighthouse upon a situation so very eligible as the Calf of Man, being the southmost point of that island."

The report was illustrated with plans of Douglas, Milford, Longships, and Portland Lighthouses. The somewhat formidable journey he had undertaken, involving 2500 miles of travelling, occupied eight weeks in its performance, and the following amusing incident shows what peaceful travellers, in those troubled times, had sometimes to encounter :—

B

"I left the Scilly Islands considerably instructed by the examination of the machinery and apparatus of this lighthouse, and very much gratified. I took my passage in a vessel bound for Penzance, where, however, I had not been long landed, when I met with a circumstance which, while it lasted, was highly disagreeable, and as it is somewhat connected with the object of the journey, I beg your indulgence while I lay it before you.

"Finding that I could not get any convenient mode of conveyance from Penzance to the Lizard Lights, I set off on foot for Marazion, a town at the head of Mounts Bay, where I was in hopes of getting a boat to freight. I had just got that length, and was making the necessary inquiry, when a young man, accompanied by several idle-looking fellows, came up to me, and in a hasty tone said, ' Sir, in the King's name I seize your person and papers.' To which I replied that I should be glad to see his authority, and know the reason of an address so abrupt. He told me the want of time prevented his taking regular steps, but that it would be necessary for me to return to Penzance, and there undergo an examination, as I was suspected of being a French spy. Had I not been extremely anxious to get on my journey, I would not have objected to this. I therefore proposed to submit my papers to the examination of the nearest Justice of Peace, who was immediately applied to and came to the inn where I was. He seemed to be greatly agitated, and quite at a loss how to proceed. The complaint preferred against me was, 'That I had examined the Longships Lighthouse with the most minute attention, and was no less particular in my inquiries at the keepers of the lighthouse regarding the sunk rocks lying off the Land's End, with the sets of the currents and tides along the coast : that I seemed particularly to regret the situation of the rocks called the Seven Stones, and the loss of a beacon which the Trinity Board had caused to be fixed upon the Wolf Rock : that I had taken notes of the bearings of several sunk rocks, and a drawing of the lighthouse and of Cape Cornwall : further, that I had refused the honour of Lord Edgecombe's invitation to dinner, who happened to be at the Land's End with a party of pleasure, offering as an apology that I had some particular business on hand, upon which I immediately set off for the Scilly Islands. These circumstances concurring with a report that a schooner had been seen off the Land taking soundings, it was presumed that I was connected with her, and had some evil intention in making these remarks.'

" In order to clear myself of this suspicion, I laid before the Justice your letter directing me to make the journey, which was signed by Mr. Gray (Secretary to the Board), as also several letters he had procured for me to some of the members of the Trinity House, London, together with a letter from the Trinity House, Leith, to the Marquis of Titchfield. I produced also my letter of credit from Sir William Forbes and Company, and, after perusing these letters, the Justice of Peace very gravely observed that they were ' merely bits of paper,' and was of opinion that I should be kept in custody till the matter should be laid before Lord Edgecombe, the Lord-Lieutenant of the county, and added, that he would most likely order me to be *sent* to Plymouth.

" I no sooner heard the opinion of this gentleman than I ordered a chaise and immediately returned to Penzance, where I laid my papers before the Justices of Peace, and waited their decision with much anxiety. They no sooner looked them over than in the most polite manner they cleared me of the suspicions I laboured under, and left me at liberty to pursue my journey, which I did with so much eagerness that I gave the two coal lights upon the Lizard Point only a very transient look, and passed on to Plymouth."

CHAPTER II.

FROM what has been said in the preceding chapter,
it will be seen that Mr. Stevenson, from an early period,
evinced a decided liking for general Engineering, and I
find that almost simultaneously with his appointment
under the Lighthouse Board, for whose peculiar duties
he had qualified himself by a pretty large and hard-
earned experience, he resolved to prosecute the practice
of Civil Engineering, in all its branches.

I find also that coincident with this start in life, he
commenced a systematic " Journal," beginning in 1801,
of the various travels made in the prosecution of his
profession, which occupies nineteen octavo and quarto
manuscript books.

His Reports, many of them on subjects of great
interest, occupy fourteen folio manuscript volumes, and
his printed reports occupy four thick quarto volumes.

These books, together with relative plans, the number of which I fear to mention, are the documents I had to consult in obtaining the records of my father's professional life. The Journals, Reports, and Plans extend over a period of nearly fifty years, and the selection of topics from such a mass of matter has been no easy task. But as the duty I have undertaken is to convey to the reader a sketch of my father as a Civil Engineer, I have been content, passing over many interesting subjects, to select from the documents before me only so much as should be useful in carrying out that object; and even in this I encountered the difficulty of determining the best order in which the selections I have made should be given. To do so according to any chronological arrangement I find to be impossible, and having resolved to give them not as a consecutive narrative, but in the form of detached notices, I think it will be most appropriate that I should commence the story of Mr. Stevenson's professional life with his great work—the Bell Rock Lighthouse,—which extended over a period of twelve years, commencing with his early conception of its structure in 1799, and terminating with its completion in 1811.

The Inchcape or Bell Rock lies off the east coast of Scotland, nearly abreast of the entrance to the Firth of Tay, at a distance of eleven miles from Arbroath, the nearest point of the mainland. The name of "Bell" has its origin in the legend respecting the good intention of a pious Abbot of Aberbrothock being frustrated by the notorious pirate, Sir Ralph the Rover, as related in

Southey's well-known lines, which I have given in an Appendix.

Of the origin, progress, and completion of the lighthouse Mr. Stevenson has left a lasting memorial and most interesting narrative in his quarto volume of upwards of 500 pages, a great part of which was written to his dictation by his only daughter, and was published in 1824.[1]

But there are some circumstances connected with the early history of the Bell Rock, which, while they could not properly have found a place in his narrative, have been noticed in his Memoranda, from which I shall transcribe a few paragraphs detailing his early efforts and disappointments while engaged in designing and arranging for the prosecution of that great work :—

"All knew the difficulties of the erection of the Eddystone Lighthouse, and the casualties to which that edifice had been liable ; and in comparing the two situations, it was generally remarked that the Eddystone was barely covered by the tide at *high water*, while the Bell Rock was barely uncovered at *low water*.

"I had much to contend with in the then limited state of my experience ; and I had in various ways to bear up against public opinion as well as against interested parties. I was in this state of things, however, greatly supported, and I would even say often comforted, by Mr. Clerk of Eldin, author of the System of Breaking the Line in Naval Tactics. Mr. Clerk took great interest in my models, and spoke much of them in scientific circles. He carried men of science and eminent strangers to the model-room which I had provided in Merchants Hall, of which he sometimes carried the key, both when I was at home and while I was abroad.

[1] Account of the Bell Rock Lighthouse. Drawn up by desire of the Commissioners of the Northern Lighthouses, by Robert Stevenson. Edinburgh, 1824.

He introduced me to Lord Webb Seymour, to Admiral Lord Duncan, and to Professors Robison and Playfair, and others. Mr. Clerk had been personally known to Smeaton, and used occasionally to speak of him to me."

It is impossible to read this little narrative without feeling a respect for Mr. Clerk's hearty enthusiasm, and perceiving the beneficial influence which a kindly disposition may produce on the pursuits of a young man, by stimulating an honourable emulation and discouraging a desponding spirit.

"But at length," the memorandum continues, "all difficulties with the public, as well as with the better informed few, were dispelled by the fatal effects of a dreadful storm from the N.E., which occurred in December 1799, when it was ascertained that no fewer than seventy sail of vessels were stranded or lost, with many of their crews, upon the coast of Scotland alone! Many of them, it was not doubted, might have found a safe asylum in the Firth of Forth, had there been a lighthouse upon the Bell Rock, on which, indeed, it was generally believed the 'York,' of 74 guns, with all hands, perished, none being left to tell the tale! The coast for many miles exhibited portions of that fine ship. There was now, therefore, but one voice,— There must be a lighthouse erected on the Bell Rock.'

"Previous to this dreadful storm I had prepared my pillar-formed model, a section of which is shown in Plate VII. of the 'Account of the Bell Rock Lighthouse.' Early in the year 1800, I, for the first time, landed on the rock to see the application of my pillar-formed model to the situation for which it was designed and made.

"On this occasion I was accompanied by my friend Mr. James Haldane, architect, whose pupil I had been for architectural drawing.   Our landing was at low water of a spring-tide, when a good *space* of rock was above water, and then the realities of its danger were amply exemplified by the numerous relics which were found in its crevices, such as a ship's marking-iron, a piece of a kedge-anchor, and a cabin stove, a bayonet, cannon-ball, silver shoe-buckle, crowbars, pieces of money, and other evidences of recent shipwreck.

"I had no sooner set foot upon the rock than I laid aside all idea of a pillar-formed structure, fully convinced that a building on similar principles with the Eddystone would be found practicable.

"On my return from this visit to the rock, I immediately set to work in good earnest, with a design of a stone light-house, and modelled it. I accompanied this design with a report or memorial to the Lighthouse Board. The abandoned pillar-formed plan I estimated at £15,000, and the stone build-ing at £42,685, 8s.   But still I found that I had not made much impression on the Board on the score of expense, for they feared it would cost much more than forty or fifty thousand pounds."

It was as to some of the details of this stone design that my father asked Professor Playfair to give his opinion, and received the following reply, which was not a little encouraging to the young engineer attempting to improve on the design of the great Smeaton :—

"Mr. Playfair is very sorry that he has scarce had any time to look more particularly over the plans which Mr. Stevenson has been so good as to send him. Mr. Playfair is too little acquainted with practical mechanics to make his opinion of much weight on such a subject as the construction of a lighthouse. But so far as he can presume to judge, the method of connecting the stones proposed by Mr. Stevenson is likely to prove perfectly secure, and has the advantage of being more easily constructed than Mr. Smeaton's."

"*9th August* 1802."

The Lord Advocate Hope, one of the Commissioners of Northern Lighthouses, and Member of Parliament for the city of Edinburgh, who had interested himself much in the Bell Rock question, and often conferred with Mr. Stevenson on his design for the work, determined that the matter should not be allowed to rest, and introduced a Bill into Parliament in 1802-1803 to empower the Board to carry it out.

This Bill passed the House of Commons. The Committee to which it was referred report—"That it appears that a sufficient foundation might be prepared on the north end of the rock, where the surface is highest and of greatest dimensions : That artificers could work five hours at the times of each low-water in the day-time of the summer months, and that if the building should be made of masonry the stones to form it might be prepared on shore, marked and numbered, and carried off to the rock and properly placed : That as the present duties may not for a long time enable the Commissioners to defray the expense of erecting and maintaining a lighthouse on the Bell or Cape Rock, it will be expedient to authorise the Commissioners to levy and take further duties for that purpose, with power to borrow a further sum on the credit of said duties."

C

At that early date there was no "standing order" of the House requiring the promoters of a Bill to lodge plans of their proposed works, and my father in his Memoranda says :—"The only plans in Mr. Hope's hands were those which, in 1800, I submitted to the Lighthouse Board."

In the House of Lords the Bill met with opposition from the Corporation of the City of London, as including too great a range of coast in the collection of duties, and such alterations and amendments were introduced in the Upper House as rendered it necessary for the Lord Advocate to withdraw the Bill.

In order to fortify Mr. Stevenson's views as to the practicability of building a stone tower in such a situation, which was apparently the chief difficulty in all the early negotiations, the Board resolved to take the advice of Mr. Telford, then employed by Government in reporting on the Highland Roads and Bridges and the Caledonian Canal, who, however, was unable to overtake the duty, and thereafter, on Mr. Stevenson's suggestion, they applied to Mr. John Rennie, Mr. Stevenson's senior by eleven years, who had, like himself, at the early age of twenty-one, commenced the practice of his profession, and was then settled in London as a civil engineer. Rennie having concurred with Stevenson as to the practicability and expediency of adopting a stone tower, the Lighthouse Board resolved to make another application to Parliament.

The second application was made in 1806, in a Bill introduced by Lord Advocate Erskine, and proceeded on the same design and estimate of £42,685, 8s., prepared by Mr. Stevenson, in 1800; and the following is an extract from the Report of the Committee of the House of Commons to whom was referred the petition of the Commissioners of the Northern Lighthouses :—

"Proceeded to examine Mr. Robert Stevenson, Civil Engineer, who, in his capacity of Engineer for the Northern Lighthouses, has erected six lighthouses in the northern parts of the kingdom, and has made the erection of a lighthouse on the Cape or Bell Rock more particularly his study,—especially since the loss of about seventy sail of vessels in a storm which happened upon the coast in the month of December 1799, by which numerous ships were driven from their course along the shore, and from their moorings in Yarmouth Roads, and other places of anchorage, southward of the Firth of Forth, and wrecked upon the eastern coast of Scotland, as referred to in the report made to this House in the month of July 1803; the particulars of which he also confirms : That the Bell Rock is most dangerously situated, lying in a track which is annually navigated by no less than about 700,000 tons of shipping, besides his Majesty's ships of war and revenue cutters : That its place is not easily ascertained, even by persons well acquainted with the coast, being covered by the sea about half-flood, and the landmarks, by which its position is ascertained, being from twelve to twenty miles distant from the site of danger.

"That from the inquiries he made at the time the

'York' man-of-war was lost, and pieces of her wreck having drifted ashore upon the opposite and neighbouring coast, and from an attentive consideration of the circumstances which attend the wreck of ships of such dimensions, he thinks it probable that the 'York' must have struck upon the Bell Rock, drifted off, and afterwards sunk in deep water: That he is well acquainted with the situation of the Bell Rock, the yacht belonging to the Lighthouse service having, on one occasion, been anchored near it for five days, when he had an opportunity of landing upon it every tide: That he has visited most of the lighthouses on the coast of England, Wales, and Ireland, particularly those of the Eddystone, the Smalls, and the Kilwarlin, or South Rock, which are built in situations somewhat similar to the Bell Rock: That at high water there is a greater depth on the Bell Rock than on any of these, by several feet; and he is therefore fully of opinion, that a building of stone, upon the principles of the Eddystone Lighthouse, is alone suitable to the peculiar circumstances which attend this rock, and has reported his opinion accordingly to the Commissioners of the Northern Lighthouses as far back as the year 1800; and having given the subject all the attention in his power, he has estimated the expense of erecting a building of stone upon it at the sum of £42,685, 8s.

"Your Committee likewise examined Mr. John Rennie, Civil Engineer, who, since the report made to this House in 1803, has visited the Bell Rock, who confirms the particulars in said report, and entertains no doubt of the practicability of erecting a lighthouse on that rock, is decidedly of opinion that a stone lighthouse will be the most durable and effectual, and indeed the only kind of building that is suited to this situation : That he has computed the expense of such a building, and after making every allowance for contingencies, from his own experience of works in the sea, it appears to him that the estimate or expense will amount to £41,843, 15s."

This application was fortunately successful, the Act having obtained the royal assent in July 1806, when the Commissioners at once determined to commence the work.

Mr. Stevenson now began to feel the full stress of his responsibility. He accordingly says in his notes :—

"The erection of a lighthouse on a rock about twelve miles from land, and so low in the water that the foundation-course must be at least on a level with the lowest tide, was an enterprise so full of uncertainty and hazard that it could not fail to press on my mind. I felt regret that I had not had the opportunity of a greater range of practice to fit me for such an undertaking. But I was fortified by an expression of my friend Mr. Clerk, in one of our conversations upon its difficulties. 'This work,' said he, ' is unique, and can be little forwarded by experience of ordinary masonic operations. In this case Smeaton's Narrative must be the text-book, and energy and perseverance *the pratique.*'"

Mr. Rennie also, who had supported the Bill of 1806 in Parliament, and afterwards was appointed by the Commissioners as an advising Engineer to whom Mr. Stevenson could refer in case of emergency, and who had suggested some alterations on Mr. Stevenson's design of the lighthouse in which he did not see his way to acquiesce, nevertheless continued to take a kind interest in the work, and they continued to correspond frequently during its progress. "Poor old fellow," Rennie says in one letter, alluding to the name of Smeaton, "I hope he will now and then take a peep of us, and inspire you with fortitude and courage to brave all difficulties and all dangers, to accomplish a work which will, if successful, immortalise you in the annals of fame."[1]

How well Mr. Stevenson met the demands which, in the course of his great enterprise, were made on his perseverance, fortitude, and self-denial, the history of the operations, and their successful completion, abundantly show. The work was indeed, in all respects, peculiarly suited to his tastes and habits; and Mr. Clerk truly— although perhaps unconsciously—characterised the man, in his terse statement of what would be required of him : "The work is unique—ordinary experience can do little for it—all must depend on energy and perseverance." No one can read Mr. Stevenson's "Account of the Bell Rock Lighthouse" without perceiving the justness of this estimate of the difficulties that lay before him, and his ability to overcome them.

Though ever maintaining the highest respect for Smeaton and his noble work, Mr. Stevenson was led, in his original design of 1800, as we have already seen, and further in his actual execution of the Bell Rock tower, to deviate to a considerable extent from the design of the Eddystone. Mr. Stevenson adopted a

[1] 7th September 1807.

height of one hundred feet instead of sixty-eight for the height of the masonry, and he carried the level of the solid part of the tower to the height of twenty-one feet above high water, instead of eleven feet as at the Eddystone. In addition to these deviations in the general dimensions of the tower, he increased the thickness of the walls, and he also introduced some changes of importance in its interior structure, whereby he secured a greater continuity, and therefore greater strength of the masonry of the walls and floors, which he describes in his book as follows :—

"Each floor stone forms part of the outward walls, extending inwards to a centre stone, independently of which they are connected by means of copper bats, with a view to preserve their square form at the extremity, instead of dovetailing. These stones are also modelled with joggles, sidewise, upon the principles of the common floor, termed feathering in carpentry, and also with dovetailed joggles across the joints, where they form part of the outward wall. . . . The floors of the Eddystone Lighthouse, on the contrary, were constructed of an arch form, and the haunches of the arches bound with chains to prevent their pressing outward, to the injury of the walls. In this, Mr. Smeaton followed the construction of the Dome of St Paul's; and this mode might also be found necessary at the Eddystone, from the want of stones in one length, to form the outward wall and floor, in the then state of the granite quarries of Cornwall. At Mylnefield Quarry, however, there was no difficulty in procuring stones of the requisite dimensions; and the writer foresaw many advantages that would arise from having the stones of the floors to form part of the outward walls, without introducing the system of arching."

Smeaton in fact adopted an arched form for the floors

of his building, which rendered it necessary, in order to counteract the outward thrust, to insert chains, embedded in grooves, cut in the masonry; but Mr. Stevenson, in designing the Bell Rock Lighthouse, improved on Smeaton's plan, not only by a better general arrangement of the masonry, but by converting the floors into effective bonds, so that, instead of exerting an outward thrust, they actually tie or bind the walls together. This is at once apparent from Figs. 1 and 2, which show the floor-courses of the Eddystone and Bell Rock in section.

FIG. 1.—Eddystone.

FIG. 2.—Bell Rock.

The engineer of the Bell Rock had all the advantage of Smeaton's earlier experience, which he ever thankfully acknowledged; but there can be no doubt whatever that the Bell Rock presented peculiar engineering difficulties. The Eddystone Rock is barely *covered* by the tide at high water, while the Bell Rock is barely *uncovered at* LOW WATER, rendering the time of working on it, as we shall afterwards find, extremely limited; and the proposal to erect a stone tower on this low-lying isolated reef, at a distance of twelve miles from land, was no less remarkable for its novelty than for its boldness.

PLATE. I.

Scale of Feet

BELL ROCK LIGHT HOUSE.

W & A.K Johnson Edinburgh.

PLATE. I.

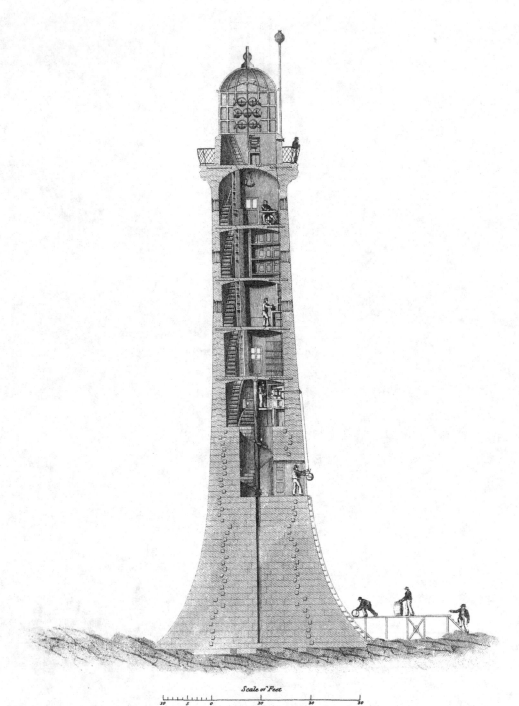

Scale of Feet

SECTION OF THE BELL ROCK LIGHT HOUSE.

Plate I. is an elevation of the Bell Rock Lighthouse, and Plate II. is a section showing the manner in which the interior is laid out, and, so far as the size of scale admits, the peculiar arrangements of the masonry, to which reference has been made.

The following is a brief statement of the progress of the work :—

The spring of 1807 was occupied in preparing a floating lightship to be moored off the rock, erecting the timber framework which was to support the barrack to be occupied as a temporary dwelling by the workmen, and in carrying out other preliminary arrangements. During this first season the aggregate time of low-water work, caught by snatches of an hour or two at a tide, amounted to no more than thirteen and a half days' work of ten hours each.

In 1808 the foundation-pit was excavated in the solid rock, and the building was brought up to the level of the surrounding surface, the aggregate time of low-water work amounting to twenty-two days of ten hours, so that little more than a month's work was obtained during the first two years.

In 1809 the barrack for the workmen was completed, and the building of the tower brought to the height of seventeen feet above high water of spring-tides.

In 1810 the masonry of the tower was finished and the lantern erected in its place, and the light was exhibited on 1st February 1811. The light is of the description known as revolving *red* and *white*, and hence Sir Walter Scott's " gem of changeful light" (see page 47).

These weary years of toil and peril were also years of great professional responsibility for the Engineer, and of constant anxiety for the safety of his devoted band of associates, including shipmasters, landing-masters, foremen, and workmen, in all of whom Mr. Stevenson took a cordial and ever friendly interest, and in whom he invariably placed implicit confidence when he found that their several duties were faithfully discharged. To form strong attachments to trustworthy fellow-workmen was ever a marked feature in my father's character, and after a lapse of nearly half a century many who joined in his labours at the Bell Rock were still associated with him in the business of his office, or as Inspectors of works.

His daily cheerful participation in all the toils and hazards which were, for two seasons, endured in the floating lightship, and afterwards in the timber house or barrack, over which the waves broke with very great force, and caused a most alarming *twisting* movement of its main supports, were proofs not merely of calm and enduring courage, but of great self-denial and enthusiastic devotion to his calling. On some occasions his fortitude and presence of mind were most severely tried, and well they stood the test.

The record of this great work is, as I have already said, fully given in the "Account of the Bell Rock Lighthouse," to which I must refer professional readers; but as this volume is out of print, and is not easily accessible, I shall give a few extracts from it, which I feel sure will be read with deep interest, and convey to the reader at least some idea of the difficulties with which this undertaking was beset :—

"Soon after the artificers landed on the rock they commenced work ; but the wind coming to blow hard, the Smeaton's[1] boat and crew, who had brought their complement of eight men to the rock, went off to examine her riding-ropes, and see that they were in proper order. The boat had no sooner reached the vessel than she went adrift, carrying the boat along with her ; and both had even got to a considerable distance before this situation of things was observed, every one being so intent upon his own particular duty that the boat had not been seen leaving the rock. As it blew hard, the crew, with much difficulty, set the mainsail upon the Smeaton, with a view to work her up to the buoy, and again lay hold of the moorings. By the time that she was got round to make a tack towards the rock, she had drifted at least three miles to leeward, with the praam boat astern ; and having both the wind and tide against her, the writer perceived, with no little anxiety, that she could not possibly return to the rock till long after its being overflowed ; for, owing to the anomaly of the tides, formerly noticed, the Bell Rock is completely under water before the ebb abates to the offing.

"In this perilous predicament, indeed, he found himself placed between hope and despair ; but certainly the latter was by much the most predominant feeling of his mind,—situate upon a sunken rock, in the middle of the ocean, which, in the progress of the flood-tide, was to be laid under water to the depth of at least twelve feet in a stormy sea. There were this morning in all thirty-two persons on the rock, with only two boats, whose complement, even in good weather, did not exceed twenty-four sitters ; but to row to the floating light with so much wind, and in so heavy a sea, a complement of eight men for each

---

[1] The tender was named after the great Engineer.

boat was as much as could with propriety be attempted, so that in this way about one-half of our number was unprovided for. Under these circumstances, had the writer ventured to despatch one of the boats, in expectation of either working the Smeaton sooner up towards the rock, or in hopes of getting her boat brought to our assistance, this must have given an immediate alarm to the artificers, each of whom would have insisted upon taking to his own boat, and leaving the eight artificers belonging to the Smeaton to their chance. Of course, a scuffle might have ensued, and it is hard to say, in the ardour of men contending for life, where it might have ended. It has even been hinted to the writer that a party of the *pickmen* were determined to keep exclusively to their own boat against all hazards.

"The unfortunate circumstance of the Smeaton and her boat having drifted was, for a considerable time, only known to the writer, and to the landing-master, who removed to the further point of the rock, where he kept his eye steadily upon the progress of the vessel. While the artificers were at work, chiefly in sitting or kneeling postures, excavating the rock, or boring with the jumpers, and while their numerous hammers, and the sound of the smith's anvil, continued, the situation of things did not appear so awful. In this state of suspense, with almost certain destruction at hand, the water began to rise upon those who were at work on the lower parts of the sites of the beacon and lighthouse. From the run of sea upon the rock, the forge-fire was also sooner extinguished this morning than usual, and the volumes of smoke having ceased, objects in every direction became visible from all parts of the rock. After having had about three hours' work, the men began, pretty generally, to make towards their respective boats for their jackets and stockings, when to their astonishment, instead of three they found only two boats, the third being adrift with the Smeaton. Not a word was uttered by any one, but all appeared to be silently calculating their numbers, and looking to each other with evident marks of perplexity depicted in their countenances. The landing-master, conceiving that blame might be attached to him for allowing the boat to leave the rock, still kept at a distance. At this critical moment the author was standing upon an elevated part of Smith's Ledge, where he endeavoured to mark the progress of the Smeaton, not a little surprised that the crew did not cut the praam adrift, which greatly retarded her way, and amazed that some effort was not making to bring at least the boat, and attempt our relief. The workmen looked steadfastly upon the writer, and turned occasionally towards the vessel, still far to leeward. All this passed in the most perfect silence, and the melancholy solemnity of the group made an impression never to be effaced from his mind.

"The writer had all along been considering various schemes— providing the men could be kept under command—which might be put in practice for the general safety, in hopes that the Smeaton might be able to pick up the boats to leeward, when they were obliged to leave the rock. He was, accordingly, about to address the artificers on the perilous nature of their circumstances, and to propose that all hands should unstrip their upper clothing when the higher parts of the rock were laid under water; that the seamen should remove every unnecessary weight and encumbrance from the boats; that a specified number of men should go into each boat, and that the remainder should hang by the gunwales, while the boats were to be rowed gently towards the Smeaton, as the course to the Pharos or floating light lay rather to windward of the rock. But when he attempted to speak, his mouth was so parched that his tongue refused utterance, and he now learned by experience that the saliva is as necessary as the tongue itself for speech. He then turned to one of the pools on the rock and lapped a little water, which produced an immediate relief. But what was his happiness when, on rising from this unpleasant beverage, some one called out 'A boat! a boat!' and on looking around, at no great distance, a large boat was seen through the haze making towards the rock. This at once enlivened and rejoiced every heart. The timeous visitor proved to be James Spink, the Bell Rock pilot, who had come express from Arbroath with letters. Spink had for some time seen the Smeaton, and had even supposed, from the state of the weather, that all hands were on board of her, till he approached more nearly and observed people upon the rock. Upon this fortunate change of circumstances sixteen of the artificers were sent at two trips in one of the boats, with instructions for Spink to proceed with them to the floating light.[1] This being accomplished, the remaining sixteen followed in the two boats belonging to the service of the rock. Every one felt the most perfect happiness at leaving the Bell Rock this morning, though a very hard and even dangerous passage to the floating light still awaited us, as the wind by this time had increased to a pretty hard gale, accompanied with a considerable swell of sea. The boats left the rock about nine, but did not reach the vessel till twelve o'clock noon, after a most disagreeable and fatiguing passage of three hours. Every one was as completely drenched in water as if he had been dragged astern of the boats."

After this accident difficulty was experienced in getting the men to turn out next morning, as related in the following extract :—

"The bell rung this morning at five o'clock, but the writer must acknowledge, from the circumstances of yesterday, that its sound was extremely unwelcome. This appears also to have been the feeling of the artificers, for when they came to be mustered, out of twenty-six, only eight, besides the foreman and seamen, appeared upon deck, to accompany the writer to the rock. Such are the baneful effects of anything like misfortune

---

[1] Spink's boat was too large to come close to the rock.

or accident connected with a work of this description. The use of argument to persuade the men to embark, in cases of this kind, would have been out of place, as it is not only discomfort, or even the risk of the loss of a limb, but life itself, that becomes the question. The boats, notwithstanding the thinness of our ranks, left the vessel at half-past five. The rough weather of yesterday having proved but a summer's gale, the wind came to-day in gentle breezes, yet the atmosphere being cloudy, it had not a very favourable appearance. The boats reached the rock at six A.M., and the eight artificers who landed were employed in clearing out the bat-holes for the beacon-house, and had a prosperous tide of four hours' work, being the longest yet experienced by half an hour.

"The boats left the rock again at ten o'clock, and the weather having cleared up, as we drew near the vessel, the eighteen artificers who remained on board were observed upon deck, but as the boats approached they sought their way below, being quite ashamed of their conduct. This was the only instance of refusal to go to the rock which occurred during the whole progress of the work."

The state of suffering and discomfort, as well as danger, on board the floating light, which lay moored off the rock during the first two seasons of the work, before the timber beacon was used as a habitation, is described in the following passage, which presents a striking illustration of the continual anxiety that must have existed in the minds of those engaged in the work, and of the frequent calls for energetic and courageous exertion :—

"Although the weather would have admitted of a landing this evening, yet the swell of the sea, observable in the morning, still continued to increase. It was so far fortunate that a landing was not attempted, for at eight o'clock the wind shifted to E.S.E., and at ten it had become a hard gale, when fifty fathoms of the floating-light's hempen cable were veered out. The gale still increasing, the ship rolled and laboured excessively, and at mid-night eighty fathoms of cable were veered out; while the sea continued to strike the vessel with a degree of force which had not before been experienced.

"During the last night there was little rest on board of the Pharos, and daylight, though anxiously wished for, brought no relief, as the gale continued with unabated violence. The sea struck so hard upon the vessel's bows that it rose in great quantities, or in 'green seas,' as the sailors termed it, which were carried by the wind as far aft as the quarter-deck, and not unfre-quently over the stern of the ship altogether. It fell occasionally so heavily on the skylight of the writer's cabin, though so far aft as to be within five feet of the helm, that the glass was broken to pieces before the dead-light could be got into its place, so that the water poured down in great quantities. In shutting out the water, the admission of light was prevented, and in the morning all continued in the most comfortless state of darkness. About ten o'clock A.M. the wind shifted to N.E., and blew, if possible, harder than before, and it was accompanied by a much heavier swell of sea; when it was judged advisable to give the ship more cable. In the course of the gale the part of the cable in the hause-hole had been so often shifted that nearly the whole length of one of her hempen cables, of 120 fathoms, had been veered out besides the chain-moorings. The cable, for its preservation, was also carefully "served" or wattled with pieces of canvas round the windlass, and with leather well greased in the hause-hole. In this state things remained during the whole day,—every sea which struck the vessel—and the seas followed each other in close succession—causing her to shake, and all on board occasionally to tremble. At each of these strokes of the sea the rolling and pitching of the vessel ceased for a time, and her motion was felt as if she had either broke adrift before the wind, or were in the act of sinking; but when another sea came, she ranged up against it with great force, and this became the regular intimation of our being still riding at anchor.

"About eleven o'clock, the writer, with some difficulty, got out of bed, but, in attempting to dress, he was thrown twice upon the floor, at the opposite side of the cabin. In an undressed state he made shift to get about half-way up the companion-stairs, with an intention to observe the state of the sea and of the ship upon deck, but he no sooner looked over the companion than a heavy sea struck the vessel, which fell on the quarter-deck, and rushed down-stairs into the officer's cabin, in so considerable a quantity that it was found necessary to lift one of the scuttles in the floor to let the water into the limbers of the ship, as it dashed from side to side in such a manner as to run into the lower tier of beds. Having been foiled in this attempt, and being completely wetted, he again got below and went to bed. In this state of the weather the seamen had to move about the necessary or indispensable duties of the ship, with the most cautious use both of hands and feet, while it required all the art of the landsman to keep within the precincts of his bed. The writer even found himself so much tossed about that it became necessary, in some measure, to shut himself in bed, in order to avoid being thrown to the floor. Indeed, such was the motion of the ship, that it seemed wholly impracticable to remain in any other than a lying posture. On deck the most stormy aspect presented itself, while below all was wet and comfortless.

"About two o'clock P.M. a great alarm was given throughout the ship, from the effects of a very heavy sea which struck her, and almost filled the waist, pouring down into the berths below, through every chink and crevice of the hatches and skylights. From the motion of the vessel being thus suddenly deadened or checked, and from the flowing in of the water above, it is believed there was not an individual on board who did not think, at the moment, that the vessel had foundered and was in the act of sinking. The writer could withstand this no longer, and as soon as she again began to range to the sea, he determined to make another effort to get upon deck.

"It being impossible to open any of the hatches in the fore part of the ship in communicating with the deck, the watch was changed by passing through the several berths to the companion-stair leading to the quarter-deck. The writer, therefore, made the best of his way aft, and on a second attempt to look out, he succeeded, and saw indeed an astonishing sight. The seas or waves appeared to be ten or fifteen feet in height of unbroken water, and every approaching billow seemed as if it would overwhelm our vessel, but she continued to rise upon the waves, and to fall between the seas in a very wonderful manner. It seemed to be only those seas which caught her in the act of rising which struck her with so much violence, and threw such quantities of water aft. On deck there was only one solitary individual looking out, to give the alarm in the event of the ship breaking from her moorings. The seaman on watch continued only two hours; he had no greatcoat nor overall of any kind, but was simply dressed in his ordinary jacket and trousers; his hat was tied under his chin with a napkin, and he stood aft the foremast, to which he had lashed himself with a gasket or small rope round his waist, to prevent his falling upon deck or being washed overboard. Upon deck everything that was moveable was out of sight, having either been stowed below previous to the gale, or been washed overboard. Some trifling parts of the quarter-boards were damaged by the breach of the sea, and one of the boats upon deck was about one-third full of water, the oyle-hole or drain having been accidentally stopped up, and part of the gunwale had received considerable injury. Although the previous night had been a very restless one, it had not the effect of inducing sleep in the writer's berth on the succeeding one; for having been so much tossed about in bed during the last thirty hours, he found no easy spot to turn to, and his body was all sore to the touch, which ill accorded with the unyielding materials with which his bed-place was surrounded.

"This morning about eight o'clock the writer was agreeably surprised to see the scuttle of his cabin skylight removed, and the bright rays of the sun admitted. Although the ship continued to roll excessively, and the sea was still running very high, yet the ordinary business on board seemed to be going forward on deck. It was impossible to steady a telescope so as to look minutely at the progress of the waves, and trace their breach upon the Bell Rock, but the height to which the cross-running waves rose in sprays, when they met each other, was truly grand, and the continued roar and noise of the sea was very perceptible to the ear. To estimate the height of the sprays at forty or fifty feet would surely be within the mark. Those of the workmen who were not much afflicted with sea-sickness came upon deck, and the wetness below being dried up, the cabins were again brought into a habitable state. Every one seemed to meet as if after a long absence, congratulating his neighbour upon the return of good weather. Little could be said as to the comfort of the vessel; but after riding out such a gale, no one felt the least doubt or hesitation as to the safety and good condition of her moorings. The master and mate were extremely anxious, however, to heave in the hempen cable, and see the state of the clinch or iron ring of the chain cable. But the vessel rolled at such a rate that the seamen could not possibly keep their feet at the windlass, nor work the handspokes, though it had been several times attempted since the gale took off.

"About twelve noon, however, the vessel's motion was observed to be considerably less, and the sailors were enabled to walk upon

deck with some degree of freedom. But to the astonishment of
every one it was soon discovered that the floating light was adrift!
The windlass was instantly manned, and the men soon gave out
that there was no strain upon the cable. The mizzen-sail, which
was bent for the occasional purpose of making the vessel ride
more easily to the tide, was immediately set, and the other sails
were also hoisted in a short time, when, in no small consternation,
we bore away about one mile to the south-westward of the former
station, and there let go the best bower-anchor and cable, in
twenty fathoms water, to ride until the swell of the sea should
fall, when it might be practicable to grapple for the moorings, and
find a better anchorage for the ship.

"As soon as the deck could be cleared the cable end was hove
up, which had parted at the distance of about fifty fathoms from
the chain moorings. On examining the cable, it was found to
be considerably chafed, but where the separation took place, it
appeared to be worn through, or cut shortly off. How to account
for this would be difficult, as the ground, though rough and
gravelly, did not, after much sounding, appear to contain any
irregular parts. It was therefore conjectured that the cable must
have hooked some piece of wreck, as it did not appear from the
state of the wind and tide that the vessel could have *fouled* her
anchor when she veered round with the wind, which had shifted
in the course of the night from N.E. to N.N.W.

"Be this as it may, it was a circumstance quite out of the
power of man to prevent, as, until the ship drifted, it was found
impossible to heave up the cable. But what ought to have been
the feeling of thankfulness to that Providence which regulates and
appoints the lot of man, when it is considered that if this accident
had happened during the storm, or in the night after the wind had
shifted, the floating light must inevitably have gone ashore upon
the Bell Rock. In short, it is hardly possible to conceive any
case more awfully distressing than our situation would have been,

or one more disastrous to the important undertaking in which we were engaged."

The distance at which the floating light was moored from the rock was about three miles, and the passage of the men to and from their work, and boarding the vessel in rough weather, was a source of great anxiety and danger, and is described in the following paragraphs :—

" When the tide-bell rung on board the floating light, the boats were hoisted out, and two active seamen were employed to keep them from receiving damage alongside. The floating light being very buoyant, was so quick in her motions, that when those who were about to step from her gunwale into a boat, placed themselves upon a "cleat" or step on the ship's side with the man or rail-ropes in their hands, they had often to wait for some time till a favourable opportunity occurred for stepping into the boat. While in this situation, with the vessel rolling from side to side, watching the proper time for letting go the man-ropes, it required the greatest dexterity and presence of mind to leap into the boat. One who was rather awkward would often wait a considerable period in this position: at one time his side of the ship would be so depressed that he would touch the boat to which he belonged, while the next sea would elevate him so much that he would see his comrades in the boat on the opposite side of the ship, his friends in the one boat calling to him to 'jump,' while those in the boat on the other side, as he came again and again into their view, would jocosely say—'Are you there yet ? You seem to enjoy a swing.' In this situation it was common to see a person upon each side of the ship for a length of time, waiting to quit his hold. A stranger to this sort of motion was both alarmed for the safety, and delighted with the agility, of persons leaping into the boat under those perilous circumstances. No sooner had one quitted his station on the gunwale than another occupied his place, until the whole were safely shipped."

On their return trips from the rock to the floating light, the men had a no less hazardous and trying ordeal to undergo, for Mr. Stevenson records the following as an example of the risks to which they were exposed :—

"Just as we were about to leave the rock, the wind shifted to the s.w., and from a fresh gale it became what seamen term a hard gale, or such as would have required the fisherman to take in two or three reefs in his sail. The boats being rather in a crowded state for this sort of weather, they were pulled with difficulty towards the floating light. Though the boats were handsomely built, and presented little obstruction to the wind, as those who were not pulling sat low, yet having the ebb-tide to contend with the passage was so very tedious that it required two hours of hard work before we reached the vessel.

"It is a curious fact, that the respective tides of ebb and flood are apparent upon the shore about an hour and a half sooner than at the distance of three or four miles in the offing. But what seems chiefly interesting here is, that the tides around this small sunken rock should follow exactly the same laws as on the extensive shores of the mainland. When the boats left the Bell Rock to-day, it was overflowed by the flood-tide, but the floating light did not swing round to the flood-tide for more than an hour afterwards. Under this disadvantage the boats had to struggle with the ebb-tide and a hard gale of wind, so that it was with the greatest difficulty they reached the floating light. Had this gale happened in spring-tides, when the current was strong, we must have been driven to sea in a very helpless condition.

"The boat which the writer steered was considerably behind the other, one of the masons having unluckily broken his oar. Our prospect of getting on board, of course, became doubtful, and our situation was rather perilous, as the boat shipped so much sea that it occupied two of the artificers to bale and clear her of water. When the oar gave way we were about half-a-mile from the ship, but, being fortunately to windward, we got into the wake of the floating light at about 250 fathoms astern, just as the landing-master's boat reached the vessel. He immediately *streamed* or floated a life-buoy astern, with a line which was in readiness, and by means of this useful implement, the boat was towed alongside of the floating light, where, from the rolling motion, it required no small management to get safely on board, as the men were much worn out with their exertions in pulling from the rock. On the present occasion, the crews of both boats were completely drenched with spray, and those who sat upon the bottom of the boats to bale them were sometimes pretty deep in the water, before it could be cleared out. After getting on board, all hands were allowed an extra dram, and having shifted, and got a warm and comfortable dinner, the affair, it is believed, was little more thought of."

An interesting incident, showing the constant anxiety of the chief for his men, is given in the following passage:—

"The boats left the ship at a quarter before six this morning, and landed upon the rock at seven. The water had gone off the rock sooner than was expected, for as yet the seamen were but imperfectly acquainted with its periodic appearance, and the landing-master being rather late with his signal this morning, the artificers were enabled to proceed to work without a moment's delay. The boat which the writer steered happened to be the last which approached the rock at this tide; and, in standing up in the stern, while at some distance, to see how the leading boat entered the creek, he was astonished to observe something in the form of a human figure in a reclining posture upon one of the ledges of the rock. He immediately steered the boat through a narrow entrance to the eastern harbour, with a thousand unpleasant sensations in his mind. He thought a vessel or boat must have been wrecked upon the rock during the night; and it seemed probable that the rock might be strewed with dead bodies—a spectacle which could not fail to deter the artificers from returning so freely to their work. Even one individual found in this situation would naturally cast a damp upon their minds, and, at all events, make them much more timid in their future operations. In the midst of those reveries, the boat took the ground at an improper landing-place; but, without waiting to push her off, he leapt upon the rock, and making his way hastily to the spot which had privately given him alarm, he had the satisfaction to ascertain that he had only been deceived by the peculiar situation and aspect of the smith's anvil and block, which very completely represented the appearance of a lifeless body upon the rock. The writer carefully suppressed his feelings, the simple mention of which might have had a bad effect upon the artificers, and his haste passed for an anxiety to examine the apparatus of the smith's forge, left in an unfinished state at the evening tide."

In the following words Mr. Stevenson explains his resolution to regard the operations at the Bell Rock as a work of mercy, and to continue them, when weather permitted, throughout all the seven days of the week :—

"To some it may require an apology, or at least call for an explanation, why the writer took upon himself to step aside from the established rules of society by carrying on the works of this undertaking during Sundays. Such practices are not uncommon in the dockyards and arsenals, when it is conceived that the public service requires extraordinary exertions. Surely if, under any circumstances, it is allowable to go about the ordinary labours of mankind on Sundays, that of the erection of a lighthouse upon the Bell Rock seems to be one of the most pressing calls which could

in any case occur, and carries along with it the imperious language
of necessity. When we take into consideration that, in its effects,
this work was to operate in a direct manner for the safety of many
valuable lives and much property, the beautiful and simple parables
of the Holy Scriptures, inculcating works of necessity and mercy,
must present themselves to every mind unbiassed by the trammels
of form or the influence of a distorted imagination. In this perilous work, to give up every seventh day would just have been to
protract the time a seventh part. Now, as it was generally supposed, after taking all advantages into view, that the work would
probably require seven years for its execution, such an arrangement must have extended the operation to at least eight years, and
have exposed it to additional risk and danger in all its stages. The
writer, therefore, felt little scruple in continuing the Bell Rock
works in all favourable states of the weather."

He however conducted a regular Sunday service, as
noticed in the following paragraph :—

" Having, on the previous evening, arranged matters with the
landing master as to the business of the day, the signal was rung
for all hands at half-past seven this morning. In the early state
of the spring-tides, the artificers went to the rock before break-
fast, but as the tides fell later in the day, it became necessary to
take this meal before leaving the ship. At eight o'clock all hands
were assembled on the quarter-deck for prayers, a solemnity which
was gone through in as orderly a manner as circumstances would
admit. Round the quarter-deck, when the weather permitted, the
flags of the ship were hung up as an awning or screen, forming the
quarter-deck into a distinct compartment with colours; the pendant
was also hoisted at the main-mast, and a large ensign flag was
displayed over the stern ; and, lastly, the ship's companion, or top
of the staircase, was covered with the *flag proper* of the Lighthouse
Service, on which the Bible was laid. A particular toll of the bell
called all hands to the quarter-deck, when the writer read a chapter

F

of the Bible, and, the whole ship's company being uncovered, he also read the impressive prayer composed by the Reverend Dr. Brunton, one of the ministers of Edinburgh."

So soon as a barrack of timber-work could be erected on the rock as a substitute for the floating light, it was inhabited by Mr. Stevenson and twenty-eight men. This barrack was a singular habitation, perched on a strong framework of timber, carefully designed with a view to

FIG. 3.—The Beacon or Barrack.

strength, and no less carefully put together in its place, and fixed to the rock with every appliance necessary to secure stability. The tide rose sixteen feet on it in calm weather, and in heavy seas it was exposed to the assault of every wave. Of the perils and discomforts of such a habitation the following passages give a lively picture:—

"This scene" (the sublime appearance of the waves) "he greatly enjoyed while sitting at his window. Each wave approached the

Beacon like a vast scroll unfolding, and in passing discharged a quantity of air which he not only distinctly felt, but was even sufficient to lift the leaves of a book which lay before him. . . .

"The gale continues with unabated violence to-day, and the sprays rise to a still greater height, having been carried over the masonry of the building, or about 90 feet above the level of the sea. At four o'clock this morning it was breaking into the cook's berth (on the Beacon), when he rang the alarm-bell, and all hands turned out to attend to their personal safety. The floor of the smith's or mortar gallery was now completely burst up by the force of the sea, when the whole of the deals and the remaining articles upon the floor were swept away, such as the cast-iron mortar-tubs, the iron hearth of the forge, the smith's bellows, and even his anvil, were thrown down upon the rock. The boarding of the cook-house, or story above the smith's gallery, was also partly carried away, and the brick and plaster work of the fireplace shaken and loosened. It was observed during this gale that the Beacon-house had a good deal of tremor, but none of that 'twisting motion' occasionally felt and complained of before the additional wooden struts were set up for the security of the principal beams ; but this effect had more especially disappeared ever since the attachment of the great horizontal iron bars in connection with these supports. Before the tide rose to its full height to-day, some of the artificers passed along the bridge into the lighthouse, to observe the effects of the sea upon it, and they reported that they had felt a slight tremulous motion in the building when great seas struck it in a certain direction about high-water mark. On this occasion the sprays were again observed to wet the balcony, and even to come over the parapet wall into the interior of the light-room. In this state of the weather, Captain Wilson and the crew of the 'Floating Light' were much alarmed for the safety of the artificers upon the rock, especially when they observed with a telescope that the floor of the smith's gallery had been carried away, and that the triangular cast-iron sheer-crane was broken down. It was quite impossible, however, to do anything for their relief until the gale should take off. . . .

"The writer's cabin measured not more than 4 feet 3 inches in breadth on the floor; and though, from the oblique direction of the beams of the Beacon, it widened towards the top, yet it did not admit of the full extension of his arms when he stood on the floor; while its length was little more than sufficient for suspending a cot-bed during the night, calculated for being triced up to the roof during the day, which left free room for the admission of occasional visitants. His folding-table was attached with hinges immediately under the small window of the apartment; and his books, barometer, thermometer, portmanteau, and two or three camp-stools, formed the bulk of his moveables. His diet being plain, the paraphernalia of the table were proportionally simple; though everything had the appearance of comfort, and even of neatness, the walls being covered with green cloth, formed into panels with red tape, and his bed festooned with curtains of yellow cotton stuff. If, on speculating on the abstract wants of man, in such a state of exclusion, one were reduced to a single book, the sacred volume, whether considered for the striking diversity of its story, the morality of its doctrine, or the important truths of its Gospel, would have proved by far the greatest treasure."

The Barrack was not removed immediately on the completion of the tower, and on Mr. Stevenson's first visit to the rock after the light had been established, it was with feelings of emotion that he viewed his old quarters. His Journal says—"I went up the trap and entered my own cabin with mingled thoughts of reflection upon the many anxious hours I had spent within the narrow precincts of its little walls, and here offered up thanks to God for the happy termination of this work."

Mr. Stevenson's merit as Engineer of the Bell Rock

Lighthouse does not rest in his bold conception of, and confident unshaken belief in, the possibility of executing a tower of masonry on that submerged reef, or even in his personal courage and discretion in carrying out so difficult a work, in the face of so many dangers, when he had neither "steamboat" nor "steam-crane" to call to his aid. But his mechanical skill in all the arrangements of the work was pre-eminent in bringing his labours to a successful issue. Not only did he conceive the plan of the moveable *jib* and *balance cranes*, described in a subsequent chapter—which he applied with much advantage in the erection of the tower, and the former of which is now in universal use,—but his inventive skill, ever alive to the possibility of improving on the conceptions of his great master, Smeaton, led him to introduce all those advantageous changes in the arrangements of the masonry of the tower, which have been already described, as distinguishing it from the Eddystone.

The Commissioners entertained a high sense of Stevenson's services at the Bell Rock Lighthouse; and, as many of them took a deep interest in the execution of that remarkable work, and paid occasional visits to it during its progress, they were well able to appreciate the ability and zeal with which he devoted himself to this arduous task, and they resolved, at a meeting held in the lighthouse itself—" That a bust of Mr. Robert Stevenson be obtained, and placed in the library of the Bell Rock Lighthouse, in testimony of the sense entertained by the Commissioners of his distinguished talent and indefatigable zeal in the erection of that lighthouse." A beautiful bust in marble, by Samuel Joseph, from which the frontispiece has been engraved, was accordingly placed in what is called the library, being the upper apartment of the tower.

Mr. Stevenson's interest in the Eddystone did not cease on the completion of his own work. We know that he paid at least two visits to the Eddystone after the completion of the Bell Rock. One of those visits was made in September 1813, when, by the courtesy of the Trinity House, he was accommodated with the use of the 'Eddystone' tender, and, though the weather was not very favourable, succeeded in landing on the rock and making a hasty inspection of the far-famed lighthouse.

Mr. Stevenson's last visit was made in 1818, on a voyage in the Northern Lighthouse tender, on which occasion he was favoured with a smooth sea and a low tide, and enabled to make a thorough inspection of the rock. It is important and interesting to record that this examination strongly impressed him with the *ultimate* insecurity of the structure, as appears from the following almost prophetic extract from his Journal :—

"The house seems to be in a very good state of repair, and does not appear to have sustained any injury by the lapse of time. The joints are full of cement, and the stone exhibits little appearance of decay, being granite or syenite. The rock itself upon a narrow inspection seems to be gneiss. The rock is shaken all through, and dips at a very considerable angle, perhaps one in three, towards the south-west; and being undermined on the north-east side for several feet, it must be confessed that it has rather an alarming appearance. I am not, however, of opinion that it has altered its state perhaps since the date of the erection of the tower. Since

PLATE III.

BELL ROCK LIGHTHOUSE

BELL ROCK LIGHTHOUSE.

Pharos loquitur

Far in the bosom of the deep
Oer these wild shelves my watch I keep
A ruddy gem of changeful light
Bound on the dusky brow of Night
The Seaman beds my lustre hail
And scorns to strike his timorous sail

my last visit in 1813 I am not sensible of any change upon it. On the north-east side, however, at what is called the 'Gut' landing-place, where the men sheltered themselves from the fire of Rudyerd's Lighthouse, but especially at low-water mark of spring-tides, there is a hollowing of the rock which penetrates at least to the circumference of the base of the lighthouse. I therefore conclude that when the sea runs high there is danger of this house being *upset*, after a lapse of time, when the sea and shingle have wrought away the rock to a greater extent. Nothing preserves this highly important building but the hardness of the rock and the dip of the strata, but for how long a period this may remain no one can pretend to say."

That period has at length arrived, and the Trinity House, under the advice of Mr. Douglass, their Engineer, have resolved that Smeaton's Eddystone—the engineer's long cherished object of veneration—must be renewed, and henceforth Stevenson's Bell Rock must be held as the earliest existing type of a class of bold and skilful works—still few in number—which, by converting a dark sunken danger into a source of light and safety, have saved many a ship, and cheered the heart of many a tempest-tossed sailor, as happily expressed in Sir Walter Scott's impromptu "Pharos loquitur," written in the Album of the Lighthouse, when he landed with a deputation of the Commissioners in 1814.

> " Far in the bosom of the deep
> O'er these wild shelves my watch I keep,
> A ruddy gem of changeful light,
> Bound on the dusky brow of night ;
> The seaman bids my lustre hail,
> And scorns to strike his timorous sail."

CHAPTER III.

LIGHTHOUSE ILLUMINATION.

1801—1843.

Early modes of illumination—Facet reflectors and lamps—Silvered copper reflectors and Argand lamps—Isle of May coal light—Improvements in catoptric lights—Distinctions for lighthouses invented by Mr. Stevenson, viz., flashing, intermittent, and double lights—Floating light lantern—Lighting of stage of Covent Garden Theatre—Dioptric system of lighthouse illumination.

SEEING that, for reasons stated in the last chapter, I was led to give up the idea of attempting to follow any chronological sequence in this Memoir, it may perhaps be convenient, before speaking of my father's general practice as a Civil Engineer, that I should supplement the sketch I have given of the Bell Rock Lighthouse by some account of the other important duties he performed as Engineer to the Commissioners of Northern Lighthouses—an office which, as we have seen, he held for so long a period.

The lighthouse towers of the last century, though useful as beacons by day, were after all most imperfect guides by night. Indeed, the rude expedients adopted at that early period to give light to the sailor in a dark and moonless sky, present a very curious contrast to the modern system of lighthouse illumination—the result of careful study by modern philosophers and engineers. If proof of this be wanted, we have only to refer to the twenty-four miserable candles, unaided by reflectors or any other optical contrivance, which shed their dim and uncertain light from Smeaton's famous Eddystone for nearly half a century after it was built.

But indeed at that early time all lights had not even the advantage of the glazed lantern which protected the candles of the Eddystone from the winter's blast and summer's breeze; the grand Tour de Cordouan on the coast of France was then lighted by blazing fagots of wood burned in an open chauffer, and many of the early lighthouses were open coal fires.

When Mr. Smith, however, was appointed Engineer to the Scotch Lighthouse Board, he, as has been already said, came forward as the advocate of lamps aided by reflectors, a system which he introduced at Kinnaird Head in 1787; so that the Lighthouse Board of Scotland never employed any less perfect mode of illumination. These early reflectors, which had been in use in England, consisted of small pieces or facets of common mirror glass arranged in a hollow mould and fixed in their places by plaster of Paris; but soon afterwards the facets of mirror glass, though forming good instruments for their day, and of their kind, were discarded, and the reflectors were thereafter made of copper, plated with silver, and brightly polished.

I am not in a position to say when or by whom these metallic reflectors were first introduced, or what was their exact form, the question being invested in some degree of doubt; but it was to the perfecting of these optical

G

instruments and adapting them to practical use in a lighthouse that Mr. Stevenson's attention was early directed. Thus we find him in 1805 reporting as follows :—

"The operations at the Start Point were this season begun upon Monday the 27th of May, and the lighthouse was finished upon Saturday the 17th October and the light advertised to be lighted upon the night of Wednesday the 1st of January 1806. Some nights before I left Sanday I had the light set in motion, when the effect appeared to be most excellent; indeed, it must be equal to the Scilly or Cromer lights, and superior to the revolving light at Tinmouth: at the former there are twenty-one reflectors, and at the latter there are fifteen, whereas at the Start Point Lighthouse I only use seven reflectors, but by altering the motion of the machinery and construction of the revolving part, I produce the desired effect."

And again in 1806 :—

"I was late in the season for making all the observations I could have wished upon the Start Point and North Ronaldsay lights, and was not very well appointed in a vessel for keeping the sea in bad weather. I however made a cruise for this purpose, and stood towards the Fair Isle in a heavy gale of wind, with an intention to run for Shetland, but the wind shifted, and I stretched towards Copinshaw, at the distance of about ten or twelve miles to the westward of Orkney, with both lights in view. The second night I went through North Ronaldsay Firth to have a west view of the lights. I put about off Westra, and stood northward with both lights in view, when it came to blow with great violence from the s.w., and it was with much difficulty we could regain the coast. Although on this trip I had rather bad weather, with a heavy swell of sea, yet it was very answerable for my purpose, and I was upon the whole much pleased with the appearance of the new light; but I find, when at the distance of ten or twelve miles, with the sea running high, the light is seen for rather too short a period, so that it would be proper to place other seven reflectors upon the frame at an angle of about 40° to the present reflectors, in the event of removing North Ronaldsay light."

I find from his correspondence that my father consulted Sir John Leslie, the distinguished Professor of Natural Philosophy, and Alexander Adie, the well-known optician, as to the best mode of procuring a true parabolic form for the construction of his reflectors, and having introduced a simple means of withdrawing the lamp from the reflector, his new catoptric apparatus may be said to have been completed.

The Bell Rock was the first lighthouse that was illuminated by Mr. Stevenson's improved apparatus (shown

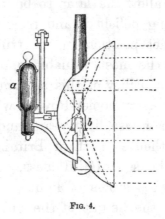

FIG. 4.

in section in Fig. 4), where a is the fountain for the oil, b the burner, and the directions of the incident and reflected rays are represented by dotted lines. In Fig. 5 the reflector is shown in elevation; the lamp is represented as lowered down from the reflector, which is

FIG. 5.

effected by a sliding arrangement controlled by a guide,—the object being to allow the lamp to be removed while the reflector is being polished, and to insure its being returned to its exact position in the true focus of the reflector. Perhaps the most valuable opinion that can be quoted as to the utility of this arrangement is that of Mr. Airy, the Astronomer-Royal, who, after the apparatus had been in use fifty years, and after having inspected the lighthouses both of Britain and France, says—"This lighthouse" (Girdleness, in Aberdeenshire) "contains two systems of lights. The lower, at about two-fifths of the height of the building, consists of thirteen parabolic reflectors of the usual form. I remarked in these, that by a simple construction, which I have not seen elsewhere, great facility is given for the withdrawal and safe return of the lamps, for adjusting the lamps, and for cleaning the mirrors;" and in closing his report he adds, "It is the best lighthouse that I have seen."[1]

Notwithstanding the introduction of this improved apparatus at the Bell Rock in 1811, a coal-fire, which had existed for the long period of 181 years on the Isle of May, at the entrance to the Firth of Forth, still continued, in 1816, to send forth its feeble and misleading light, and as it was one of the best specimens of the lighthouses of days now passed away, it may not be uninteresting to give a short account of it.

The May light was at that period what is called a "private light"—the right of levying dues on shipping being vested in the Duke of Portland, who was owner of the island. There were many private lights in England, but the Isle of May was the only one that still remained in Scotland, and the Commissioners of Northern Lighthouses, believing it to be advantageous that so important a light should be placed under public management, so as to secure for the shipping a better light, and exemption from the high passing tolls charged by the proprietor, entered into treaty with the Duke of Portland for the purchase of his rights. This negotiation resulted in the introduction of a Bill into Parliament in 1814, authorising the purchase of the Isle of May, with the right of levying toll, for the sum of £60,000.

So soon as the property came into the hands of the Commissioners they erected a new lighthouse, and on the

[1] Report of the Royal Commission on Lighthouses, 1861, p. 86.

1st of February 1816 the old coal chauffer was discontinued, and a light from oil with reflectors was exhibited in its stead. I am enabled from an old plan in my possession to present the reader with two sketches of the original chauffer light of the Isle of May.

Fig. 6 is an elevation of the building, with the tackle for raising the fuel to the top, and its inscription stone

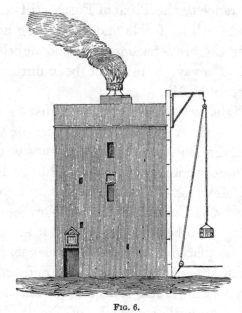

FIG. 6.

over the door bearing the date 1635. Fig. 7 shows the building in section, with its stone winding staircase and vaulted chambers, the whole structure apparently being so designed as to be perfectly proof against fire—a precaution very necessary for a building dedicated to such a purpose, for it is recorded that no fewer than 400 tons of coal were annually consumed in the open chauffer on its top.

It was, as I have said, one of the best coal-fires in the kingdom, and three men were employed to keep the bonfire burning, so that its inefficiency as a light was not due to any want of outlay in its support. But its appearance was ever varying, now shooting up in high flames, again enveloped in dense smoke, and never well seen when most required. When Mr. Stevenson visited the island, with a view to its purchase by the Commissioners, he

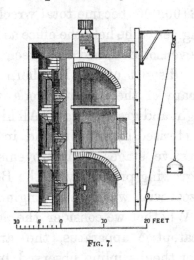

Fig. 7.

was told by the keeper, that in violent gales the fire only kindled on the *leeward* side, and that he was in the habit of putting his hand through the *windward* bars of the chauffer to steady himself while he supplied the fire with coals, so that in the direction in which it was most wanted hardly any light was visible. Nothing can be worse than any variableness or uncertainty in the appearance of a light. Better far not to exhibit it at all than to show it irregularly ; and the coal lights were so

changeable and destitute of characteristic appearance as
to be positively dangerous. This indeed was too sadly
proved by the loss of H. M. ships 'Nymphen' and
'Pallas,' which on the 19th December 1810 were wrecked
near Dunbar, the light of a limekiln, on the coast of
Haddington, having been mistaken for the coal light of
the Isle of May. Fortunately only nine of their combined crews of 600 men perished; but the vessels, valued
at not less than £100,000, became total wrecks.

During the long period he held the office as Engineer to
the Board, Mr. Stevenson designed and executed eighteen
lighthouses in the district of the Northern Lighthouse
Commissioners, many of them in situations which called
for much forethought and great energy. All his lighthouse
works were characterised by sagacity and inventiveness,
and exhibit successive stages of improvement, equally
indicative of the growing prosperity of the Board and of
the alacrity and zeal with which their Engineer laboured
in his vocation. Whether we consider the accuracy and
beauty of the catoptric apparatus, the arrangements
of the buildings, or the discipline observed by the lightkeepers of the Northern Lighthouses, we cannot fail to
recognise the impress of that energetic and comprehensive
cast of mind which directed the whole. Acting under
the direction of an enlightened Board of Commissioners,
my father may, with the strictest propriety, be said to
have created the lighthouse system of Scotland. His
merits indeed in this respect were generally acknowledged in other quarters; and many of the Irish lighthouses, and several lighthouses in our colonies, were
fitted up with apparatus prepared after his designs.

In the course of his labours my father's attention was much given to the question of *distinction* among lights—a matter of the utmost importance, especially in narrow seas, where many lights are required; and at his suggestion, the Northern Lighthouse Commissioners fitted up a temporary light-tower on Inchkeith, in which numerous experiments having this object in view were made.

He was the inventor of two useful distinctions—the *Intermittent* and *Flashing* lights. In the intermittent distinction the light is suddenly obscured by the closing of metallic shades which surround the reflector frame, and on their opening, it is as suddenly revealed to sight, in a manner which completely distinguishes it from the ordinary revolving light, which from darkness, *gradually* increases in power till it reaches its brightest phase, and

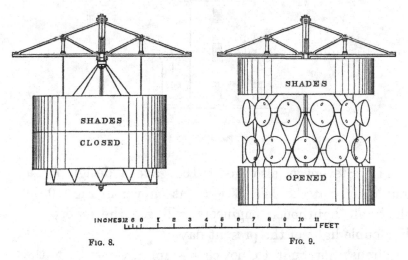

FIG. 8.   FIG. 9.

then gradually declines until it is again obscured; the action of these shades in producing the intermittent effect is illustrated in Figs. 8 and 9. The *Flashing* light, by a peculiar arrangement of reflectors, and a rapid revolution of the frame which carries them, is made to give a sudden flash of great power, once in five seconds of time, and thus has a distinctive appearance very different from either the revolving or intermittent light. For these distinctions Mr. Stevenson received from the King of the Netherlands a gold medal as a mark of his Majesty's approbation.

Fig. 10.

Mr. Stevenson also, in 1810, gave a design for a double light at the Isle of May, as shown in Fig. 10, in which all lighthouse engineers will see the embryo of the double light of the present day.

I must not omit to notice his improvement on the lanterns of floating lightships, now universally adopted, which he introduced in 1807. Previously to this date the lightships exhibited their lights from small lanterns suspended from the yardarms or frames. Mr. Stevenson realised the inutility of such a mode of exhibition, and conceived the idea of forming a lantern to surround the mast of the vessel, and to be capable of being lowered down to the deck to be trimmed, and raised when required to be exhibited. His plan had the advantage of giving a lantern of much greater size, because it encased the mast of the ship, and with this increase of size it enabled larger and more perfect apparatus to be introduced, as well as gearing for working a revolving light. Fig. 11 shows this lantern, and the following is his description of it :—

"The lanterns were so constructed as to clasp round the masts and traverse upon them. This was effected by constructing them with a tube of copper in the centre, capable of receiving the mast, through which it passed. The lanterns were first completely formed, and fitted with brass flanges; they were then cut longitudinally asunder, which conveniently admitted of their being screwed together on the masts after the vessel was fully equipped and moored at her station. Letters *a a* show part of one of the masts, *b* one of the tackle-hooks for raising and lowering the lanterns, *c c* the brass flanges with their screw-bolts, by which the body or case of the lantern was ultimately put together. There were holes in the bottom and also at the top connected with the ventilation; the collar-pieces *e* and *g* form guards against the effects of the weather. The letter *h* shows the front of the lantern, which was glazed with plate-glass; *i* is one of the glass shutters by which the lamps were trimmed, the lower half being raised slides into a groove made for its reception; *k* shows the range of ten agitable burners or lamps out of which the oil cannot be spilt by the rolling motion of the ship. Each lamp had a silvered copper reflector *l* placed behind the flame."

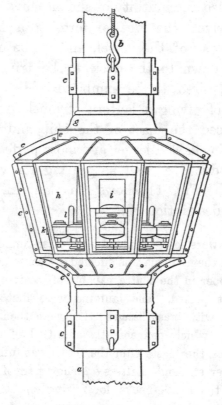

Fig. 11.

The reputation of my father's catoptric apparatus was not, it appears, confined to those interested in the welfare of the seaman. In 1819, Mr. Stevenson was waited on by a gentleman passing hurriedly through Edinburgh, who came on behalf of Mr. Harris, the manager of Covent Garden Theatre, who was desirous to try catoptric apparatus for certain stage effects

which he intended to introduce in London. The proposal
seems rather to have taken the Lighthouse Engineer by
surprise, but on learning that the gentleman who had
favoured him with a call was Mr. Benson, the famous
singer of the day, he wrote the following letter to Mr.
Harris :—

"I had some conversation with Mr. Benson of your theatre on
the day he proposed to leave this for London. The purpose of his
visit to me was to inquire about the reflectors we used in the
lighthouses upon this coast, which are under my direction, as he
had some plan in view for dispensing with the *footlights* on the
stage by the introduction of reflected light.

"Being desirous to give every facility to Mr. Benson's views,
I offered him the loan of a reflector, which I showed him; but from
his being on the eve of setting off, and wishing to keep the
discovery, if practicable, for your theatre, I agreed to send it to
you at Covent Garden, and this letter is to acquaint you that
a case containing the reflector and its burner was shipped to your
address.

"You are to understand that there is no charge whatever to be
made; I only request that the reflector may be returned when you
have made your trials. I no sooner learned that I conversed with
the gentleman who sings so delightfully in 'Rob Roy' than I felt
an irresistible inclination to oblige him.

"Wishing you every success in the projected improvement in
lighting the stage, I remain," etc.

The reflector was duly returned by Mr. Harris. The
note intimating its shipment says—"It is an excellent
reflector, but it collects the light too much in one
spot for our use; I mean, it does not spread the light
sufficiently about."

I mention this small matter, not so much because the manager of Covent Garden Theatre came to Edinburgh to get his information, but to show that Mr. Harris's experiment, made in 1819, foretold the result of all trials that have since been made to light railway stations, public gardens, and parks, by using lighthouse apparatus, which is designed to *condense* the rays of light, and not to *diffuse* them, and is therefore inapplicable for such purposes.

The remarks I have made on lighthouse illumination refer to what is known as the *catoptric* system, whereby the light is acted on by *reflection* alone. The invention of the *dioptric* system by Fresnel was first communicated to Mr. Stevenson in a letter received from Colonel Colby of the Royal Engineers, who had an opportunity of knowing the benefit of Fresnel's dioptric light in making certain trigonometrical observations for connecting the Government surveys of the shores of England and France across the English Channel. The letter is in the following terms :—

"Tower, *1st Nov.* 1821.

"My dear Sir,—I am quite ashamed of having delayed answering your letter, and thanking you for the communications you sent me for so long a time. In regard to the lamps, an account will be given of them in the *Annales de Chimie* for the next month. The lens is composed of pieces of glass forming a circle three feet in diameter, ground to three feet focal length. The lamp is similar to an Argand lamp, having hardly any other difference, except four concentric circular wicks instead of one. The external wick is about three inches in diameter. The light given by the lens is remarkably brilliant. When we were at Folkestone Hill, the lamp at Blancnez appeared to give about four times the light of the Dungeness Lighthouse, though the distance of the lamp was nearly double that of the lighthouse. The only difficulty which occurs to me in their employment in lighthouses is the small angle to which a single lens gives light. I think one lens is brilliant for seven degrees, and could not answer for more than eight or nine degrees.

"The Cordouan Lighthouse is to be fitted up with ten lenses round one lamp.

"With best wishes to Mrs. S. and your family, ever yours, Thos. Colby."

The merits of the dioptric system of illumination were brought before the Commissioners of Northern Lighthouses in Mr. Stevenson's Report of December 1821, and, as is well known, it has, with various extensions and important improvements, been very generally adopted in all cases where it is applicable to lighthouse illumination.

CHAPTER IV.

ROADS.

1798—1835.

Early roads and road-making—Edgeworth and M'Adam's systems of roads
—Stevenson's system of roads—Cast-iron and stone tracks.

WRITING at an early date, Mr. Stevenson has given the following sketch of Roads and Road-making :—

"In early periods, when every family formed a kind of community within itself for providing the necessaries of life, it is obvious that there could be little communication with distant parts of the country, and there was, therefore, no use for roads, which, long after the establishment of towns, must have continued in the state of *footpaths* and *horse-tracks*. The bulky articles of fuel and building materials are likely to have given rise to the first idea of a sledge, the precursor of the wheel-carriage, which ultimately led to the construction of anything like a regular path. The first roads of Britain appear to have been the Military Ways of the Romans. Some remains of these are still to be seen in various parts of the kingdom, and even in the immediate vicinity of the city of Edinburgh. It is, however, quite astonishing how slow the progress of improvement in road-making seems to have been, and especially its adaptation to economical purposes; although all classes must have felt an equal interest in the formation of roads, as both the landed proprietor and the citizen were to be mutually benefited by thus laying open the country. But it requires the accumulated wealth of ages to produce improvements so expensive. It is long before the mind can be brought to approve of any radical change of habit, however advantageous; and the scale adopted in the first instance is often so circumscribed, that the whole measure requires to be extended and even to be changed a second, and perhaps a third time, in keeping pace with the public demands for improvement.

"It is well known, that even so late as about the middle of the last century, almost the whole land carriage of Scotland, and a great part of England, was conducted upon horseback, the animals employed being termed *pack-horses*. To the horse-tracks thus produced, and which in the first instance were *formed* without regard to steep acclivities, are to be ascribed the evils which we now labour under, as attendant on the laying out of our roads for the modern improvement of wheel-carriages. Nor was it till after much practice and the application of scientific principles, long after the introduction of carriages, that we were induced to improve the line of draught and adopt level tracks of road, although perhaps more circuitous.

"In Great Britain the road department, after much experience, is now brought into a system by which the highways are made and upheld by dues directly levied on those who travel or use them,—excepting, indeed, such roads as are situated in very remote parts of the country, where the Government, with the most enlightened policy, has either executed the works directly by the troops upon the *peace establishment*, as in the case of General Wade's army, or given aid towards the original formation of extensive lines of road, for opening the more remote districts of the country. There is, perhaps, no better criterion for judging of the prosperity of a country than by its public improvements; and were this subject considered in all its bearings, we should hardly be able to quote any stronger evidence of internal riches and true greatness, than we find connected with the subject of its public roads. It appears from a very general or cursory calculation,

I

which the reporter has made, that the highways of Great Britain and Ireland, independently of the almost innumerable parish and private roads, extend to about 25,000 miles. The expense of these, including bridges, etc., on a very moderate calculation, may be stated throughout the kingdom at the rate of £800 per mile, which is equal to no less than the aggregate sum of twenty millions sterling. Now, to what branch of political economy can we look with more certainty and propriety than to such splendid examples of the substantial wealth and resources of a country? for until a kingdom is traversed and laid open by roads, its government must be weak, and its people remain in a state of comparative poverty.

"But in so extensive a concern as the system of roads, involving so great an expense, we may naturally look for small beginnings and very gradual advancement. Accordingly, we find in the first formation of highways, before their utility could be fully understood or experience had shown the benefits of science in the practice of the engineer, the early road-maker only increased the breadth of the horse-track, and strewed it over with gravel from the neighbouring brook. Indeed, we know that so late as the year 1542, even the streets of London were formed in this way; and it is said to be established by the records of Parliament, that when the new system of road-making was first proposed to be extended beyond the region of a few miles from that metropolis, such was the mistaken policy and narrow-minded views of the immediate proprietors, that the measure was strenuously opposed by those who wished to make a monoply of the supplies for the metropolis, as detrimental to the established order of things."

The names of Richard Edgeworth, F.R.S., and John M'Adam, are well known in connection with roads—Mr. Edgeworth writing in 1813, Mr. M'Adam in 1816. Both men had, it appears, given attention to the subject before the end of the last century. Mr. Edgeworth says :—" I have visited England, and have found, on a journey of many hundred miles, scarcely twenty miles of well-made road. In many parts of the country, and especially near London, the roads are in a shameful condition, and the pavement of London is utterly unworthy of a great metropolis."

Mr. M'Adam had been much struck by the entire want of system that existed in the management of roads at that early period, and strongly urged the necessity of a reform in road *management* as a pre-requisite to road *improvement*. He urged the laying out of the roads of the country into separate districts, with the appointment of road trustees to manage them—the appointment of chief and assistant road-surveyors to superintend them—and a new system of accounting and finance,—all under statutory regulations; and it cannot be doubted that in all this Mr. M'Adam did good service, which was recognised in 1823 by Parliament voting a sum of money to him for having introduced a system of "repairing, making, and managing turnpike roads and highways, from which the public have derived most important and valuable advantages."

It appears to me, however, that all that is said in Mr. M'Adam's first edition of his book on road-making, in 1816, is of so general and vague a nature that he cannot have known of Mr. Stevenson's work at an early part of the century.

From Mr. Stevenson's reports it appears that he was much employed in road-engineering in the counties of Edinburgh, Stirling, Linlithgow, Perth, and, indeed, generally throughout Scotland, extending as far north as Orkney and Shetland; and without raising any claim to priority of design, I give the following extracts from reports made by him in 1812 and 1813, after he must have had at least several years' previous study and practice of road-making, which I think clearly show that Mr. Stevenson, if not the *original*, was at least an *independent* inventor of the system of road-making which is termed "macadamising."

In a report to "The Honourable the Committee of the Trustees for the Highways and Roads within the county of Edinburgh," dated 1812, he says :—

"It may not, however, be considered altogether out of place to notice that the pieces of stone composing the road-metal in common use are perhaps one-half, and in some instances two-thirds, larger than is suitable for the best condition of a road. Road-metal of a small size consolidates by the pressure of weighty carriages, when stones of the size *commonly used* are either pounded under the wheel or forced into the road. It would therefore be desirable, as an experiment upon the large scale, to lay one of the most public roads in the county to the extent of one-fourth of a mile with stones broken much smaller than is *customary*.

"In some instances, especially within a few miles of Edinburgh, it might be worthy of consideration by the Honourable Trustees of this county how far *cast-iron cart-tracks* might not be advantageously laid upon the roads. Some years since the reporter got two or three yards' length of these iron tracks brought from the Shotts ironworks, where they have been used for years with much advantage, and, it is believed, with economy. These cart-tracks would cost about £2000 per statute mile, including upholding by the iron-founder for one year. It would be interesting to have also a trial made of these in some very public road, although it were only to the extent of two or three hundred yards."

Again, in a report to "The Honourable the Trustees for the Bridge of Marykirk," also in 1812, he says :—

"In the annexed specification of road-makers' work, the reporter makes some alterations upon the *common* and *ordinary* method of breaking and laying road materials, by reducing the road-metal to a more uniform size, and using a course of gravel, if it can be procured, or even of clean sharp sand, as a bottoming for the broken stones. A road composed of stones of various sizes can never be brought into that smooth and uniform surface, which is so much to be desired, for the moment the pressure is brought upon one of these *out-sized* stones, it must either be crushed under the wheel or be forced by repeated attacks into the road, and thereby it displaces the surrounding stones, and in either case admission is given to the surface-water ; a pit is immediately formed, and every succeeding wheel widens the breach, until the road is rendered impassable. To counteract this very common effect, arising chiefly from the very vague manner of defining the dimensions of road-metal by bulk or even by weight, the reporter provides that the Trustees shall furnish a riddle or screen, the meshes or openings of which are to be of such dimensions that a stone measuring more than one inch and a half upon any of its sides cannot pass through it."

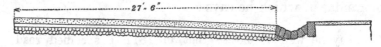

Fig. 12.—Section of one half of Roadway.

Mr. Stevenson's specification of the Regent Road in Edinburgh is fuller, and is in the following terms :—

"The cross section (shown in Fig. 12) of the metalled road to be the same in all respects as that already described for the causewayed roadway.   But the cross section is to rise from the interior *brows* or slopes of the paved channels to the centre of the roadway, at the rate of 1 in 25.   The bottoming of the road is to be of broken stones from the excavated matters of the Calton Hill works ; the pieces of stone not to exceed five or six lbs. in weight ; to be laid *by hand* in a compact manner to the depth necessary for preparing the road for the upper strata, viz., a layer or stratum of clean sharp sand four inches in thickness, laid all over the surface, and forming a bed for the upper or road-metal stratum, which is to be seven inches in thickness, and to consist of broken stones taken from the quarries of Salisbury Crags, or the lands of Heriot's Hospital, as may be finally agreed upon.   The road-metal is to be broken into pieces of such dimensions as to pass freely through a screen, to be provided by the Commissioners, the meshes of which shall not exceed one inch and a half square.   The whole to be finished with a 'top-dressing' of sea-gravel, in such a manner that none of the road-metal shall appear on the surface of the roadway when it is completed."

These extracts, so far as I have been able to discover, contain the earliest proposals and precise specification of the construction of road now known by the familiar name of "macadamising," and I dismiss the subject with the following candid quotation from Mr. Stevenson's Memoranda, in which he says :—

"It may be well to notice that in 1811 I specified road materials of the size as nearly as may be of road-metal, which *afterwards* became what is called 'macadamised roads.' I am not sure if I was before Mr. M'Adam in this respect ; at all events he had the great merit of introducing the system of smooth roads. When I first proposed this method, I think, to the Trustees of Marykirk, they objected to it upon the score of expense."

As regards the iron cart-tracks suggested for trial by Mr. Stevenson in his report to the Edinburgh Road Trustees, already quoted, he subsequently matured his views and described them in the article " Roads" in the *Edinburgh Encyclopædia*, where he proposed to use stone tracks as a "smooth and durable city road," which he describes as follows :—

"The individual component stones of the wheel-tracks, hitherto very partially in use, extend from three to four feet in length, are about ten or twelve inches in breadth, and eight or ten inches in depth. The stones of the tracks recommended by me, on the other hand, are of a cubical form, measuring only from six to eight inches in the lengthway of the track, and twelve to fourteen inches in depth, eighteen inches in breadth at the base, and twelve inches at the top or wheel-track. The stones are therefore proportionate in all their dimensions, for unless they contain a mass of matter corresponding to their length, they will be found to want strength and stability. It would hardly be possible to keep slender stone-rails in their places, and hence the chief benefit of a connected railway would be lost. On the other hand, very large materials are difficult to be got, and are also more expensive in carriage and in workmanship than stones of a smaller size. The Italian wheel-tracks are composed of stones two feet in breadth, and of various lengths. To lessen the risk of horses falling, these broad stones are kept in a rough state, by occasionally cutting grooves with a pick-axe upon their upper surface. A mode of paving with large blocks of granite, chequered or cut in this manner, has been tried in some of the streets in London. In order, however, to give pavement of this kind the necessary stability, the blocks would require to have their dimensions equally large on all sides, the expense of which would be too great. But cubical stones of the size now recommended may be procured at a moderate price, and throughout a great range of country; while the tracks, if properly laid, will actually be more stable than if blocks of larger dimensions were employed. For we may notice that a carriage-wheel rests or impinges even upon a less surface than one inch of its track at a time, in the course of each revolution round its axis; hence, it may be conceived to produce a kind of compensating effect, connected with the use of small stones, which prevents the tremor from being communicated beyond the limited sphere of each particular block, and, consequently, extending only a few inches. This system of paving I originally proposed for the main street of Linlithgow, forming part of the great western road from Edinburgh to Stirlingshire, and a correct idea of the proposal will at once be acquired by examining Fig. 13. By using tracks of this

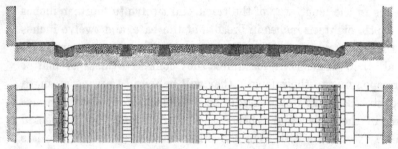

FIG. 13.—Section for Road Metal.   Section for Causeway.

description—giving the stones a proportionally broad bed, and laying them upon a firm foundation (which is indispensable)—we should have our streets and the acclivities of our highways rendered smooth and durable, avoiding the expense and inconvenience of the common road, and also the irksome noise and jolting motion of the causeway.

"The tracks may be formed of granite, whinstone, or any of the hard varieties of rock capable of being hammer-dressed."

Specimens of these stone tracks were laid in Edinburgh, in terms of Mr. Stevenson's specification, on

South Bridge Street opposite to the College, and in the Pleasance, and a third specimen was laid by the Road Trustees on Liberton Hill, which still remains after a lapse of half a century.

Subsequently to this Mr. Walker laid similar tramways in the Commercial Road, London, and as is well known, they have been pretty largely used in the principal towns in Italy.

For a "city road," as Mr. Stevenson termed it, the system he proposed has certain advantages, inasmuch as carriages with any form of wheel may use it, and this freedom of use admits of any amount of traffic being accommodated, carriages having the freedom of passing from the stone track to any part of the road. The introduction of iron "street tramways" may, however, be said, for the present, to have taken the place of all other plans for improving city passenger traffic.

CHAPTER V.

IMPROVEMENT OF EDINBURGH.

1812—1834.

Design for approaches to Edinburgh from the East by Regent and London Roads, and opening up access to the Calton Hill—Sites for the new Jail and Court of Justiciary, and buildings in Waterloo Place—Regent Bridge—Feuing Plan for Eastern District of Edinburgh—Improvement of accesses to Edinburgh from the West and North, and from Granton—Removal of old "Tolbooth" Prison—Removal of University buildings.

ANCIENT Edinburgh was famed for its narrow streets and crooked wynds, and even at the period when this Memoir begins, much remained to be done for the improvement of the various accesses to the city. These roads, leading from north, south, east, and west, were under the management of different Trusts or public bodies, by all of whom Mr. Stevenson was on various occasions consulted; and the subject seems to have had for him more than a merely professional interest, for his advice was generally far "ahead" of the cautious views of his employers, on whom he seems often to have had no small difficulty in urging the adoption of sufficiently comprehensive designs. His love for the beautiful rose above all other feelings, and he succeeded, not without difficulty and perseverance, in securing for Edinburgh those spacious road improvements which have undoubtedly helped her to claim the title of "Modern Athens."

The "Modern Athenians" who now enjoy the magnificent approach to Edinburgh by the Regent Road and Calton Hill, or that no less commodious access from Parson's Green to Leith Walk, known as the "London Road," can hardly realise the time when the only communication from Princes Street to Portobello was by Leith Street, Calton Street, and the North Back of the Canongate.

At that time Princes Street was abruptly terminated by a row of houses at the Register Office, and the Calton Hill was in a state of nature.

Mr. Stevenson's scheme of forming a direct access to London and the south, by making a roadway over the Calton Hill, was based on a comprehensive scale, providing sites for public buildings, and an extensive feuing-plan for the eastern portion of Edinburgh, all of which were ultimately carried out under his directions.

But this scheme, boldly conceived and so beneficial to Edinburgh, was not well received by the inhabitants. It had the *economical* objection of interfering to some extent with house property, a liberty to which people were only reconciled in modern times when sites had to be acquired for railway stations. It had the *engineering* objection of involving what were represented in those days as dangerous rock cuttings and extensive high retaining walls along the sides of the Calton Hill; but above all, it had the serious *social* objection that its route ran through the "Old Calton Burying-ground," and involved the removal of the remains of those interred in it to a new resting-place, to be provided by the Improvement Commissioners. This last objection subjected Mr. Stevenson to some ill feeling; and the fact that the place of interment of his own family was one of those to be removed to the new cemetery, did not succeed in allaying the discontent. It was undoubtedly in consequence of Mr. Stevenson's perseverance and unfaltering conviction that his advice was *sound*, and calculated to benefit his fellow-citizens, that his plan was ultimately adopted and carried out.

It is proper to notice that the new jail and the buildings in Waterloo Place were designed by Mr. Archibald Elliot, and at a more recent period the houses in the Regent and Royal Terraces by Mr. Playfair, and the High School and Burns's Monument by Mr. Thomas Hamilton, all architects of eminence, whose works added to the attractiveness of Mr. Stevenson's splendid access.

In carrying the road round the part of the hill now occupied by the High School, Mr. Stevenson had some difficulty, owing to the height of the retaining wall, in avoiding what would have appeared as a dead wall, and would have proved unsightly as viewed from Arthur's Seat. He accordingly built a strong retaining wall of masonry, which supports the road, and is covered by an exterior wall of rough masses of stone arranged as rustic work, which, when viewed at a distance, has all the appearance of a face of natural rock.

In Lord Cockburn's *Memorials of his Time* he says:—"Scarcely any sacrifice could be too great that removed the houses from the end of Princes Street, and made a level road to the hill, or, in other words, produced Waterloo Bridge. The effect was like drawing up the curtain of a theatre."

PLATE IV.

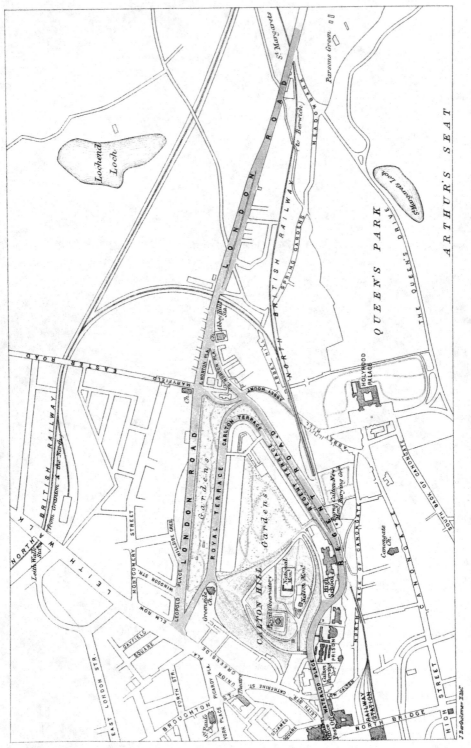

APPROACHES TO EDINBURGH BY REGENT AND LONDON ROADS, 1814.

J. Bartholomew Edin.

In Plate IV. are traced, in red colour, the various lines of connecting road which go to make up this grand improvement, of the value of which those who know the locality can judge for themselves.

In the following report, addressed to the "Sheriff-Depute of the county of Edinburgh, as convener of a committee for erecting a new jail for the county of Edinburgh," Mr. Stevenson details the various benefits to be derived by adopting his proposal; and as his views on this matter encountered, as has been stated, much opposition, I give extracts from his report, begging of those readers who have no local interest in it kindly to pass it over:—

"In the report which you addressed to the Commissioners for erecting a new jail for the county of Edinburgh, the Calton Hill is amongst other places alluded to as a site. But the difficulty of access to that commanding and healthful situation presents itself as a strong objection to its being adopted. As, however, an approach to the city from the eastward, with access to the extensive lands connected with the Calton Hill, valuable both as building-grounds and as a delightful city walk, has long been a *desideratum*, and as the present seemed a fit time for again attempting this measure, the reporter had the honour to receive your instructions to inquire into the practicability of making a proper communication to the Calton Hill, with the view of there building the intended new jail; and he is now to submit the accompanying survey of the grounds, together with the requisite plans and sections connected with the design of a road from Shakespeare Square, at the eastern extremity of Princes Street, to join the great road to London at the Abbey-hill.

"The Hon. Sheriff is aware that the attainment of this object has long been wishfully kept in view by the public. It is believed

that at different times such proposals were by them brought under
the notice of Mr. Adam and of Mr. Baxter, the most celebrated
architects of their day.   But still the work remains to be accomplished, not certainly from any physical difficulty necessarily
attending its execution, but from the want of sufficient energy to
meet the expense that must unavoidably attend an operation of
this nature, involving the removal of some valuable buildings, and
otherwise interfering with private property.   Were the reporter to
have in view merely the forming of an improved approach to the
city of Edinburgh from the eastward, instead of the present inconvenient access by the Water Gate, he might here allude to the
intended London Road through the lands of Hillside to Leith Walk,
or to the once proposed line of road terminating by a bridge from
the northern side of the Calton Hill to Greenside, opposite York
Place, and the completion of this fine street by the removal of the
old and ruinous houses which still continue to encumber its
entrance ; or he might take notice of the less commodious road at
one time in view over the higher parts of the Calton Hill, and joining the lower part of Leith Street by means of an arch over Calton
Street.   But all of these lines of road are *objectionable*, in a greater
or less degree, inasmuch as they include the acclivity of Leith
Street before the passenger can arrive at the level of the North
Bridge.   To obtain this in the most eligible manner, we must look
to the extension of the line of Princes Street to the Calton Hill,
for although the other lines of road have been looked forward to as
improvements to a certain extent, yet still they were defective, and
must have left something undone, while the extension of Princes
Street by a bridge. over Calton Street, and a road to the Abbey-
hill, seems to answer every purpose.   It unfortunately happens,
however, that if carried in a direct line it must pass through the
Calton Burying-ground ; and if this part of the road were made with
a curve, the most desirable effect in point of beauty would not be
produced.   There was a time indeed when, without encroachment

upon the burying-ground, the road could have been made with a curve to the southward of Hume the historian's tomb; but of late years the walls of the burying-ground have been extended to the verge of precipitous rocks, so that the removal of numerous private cemeteries would now be indispensable in carrying the road at an elevation sufficient to command the proper view. If a lower level were adopted in this direction, the fine prospects of the higher road would be lost, and this line would then become quite uninteresting, while a heavy expense must be incurred in carrying the road through much private property, considerations which are sufficient to render this line highly objectionable.

"But the road which would afford the easiest line of draught is that which the reporter has delineated upon the plan by a curved line towards the left from the eastern extremity of the new bridge, crossing the present road to the Calton Hill, winding round the northern side of the hill and joining the intended 'London Road' through the lands of Hillside near the eastern road to Leith. By this line of road the level of Princes Street may be conceived to become the summit level of the road, which would admit of being made with a uniform declivity from Shakespeare Square to the Abbey Hill, while the acclivity to Bridewell by the present road might be greatly reduced, and the road improved in connection with the new line of road. In the present instance, however, it is not to the easiest line of draught as an approach to the city of Edinburgh that the Sheriff directs the attention of the reporter, but to a better access to the higher lands of Calton Hill, with a view to obtain a proper site for the new jail, and therefore only an eye view of the northern line of road is given. Yet when a communication is opened with the Calton Hill by a bridge from Princes Street, we may expect at some future day to see one continuous street or drive round the hill. Before proceeding further, a preliminary remark may here be stated, and in making it the reporter thinks it proper to say that no one can hold the great professional abilities of Mr. Adam in higher estimation than he does; at any rate he is certain that it could not fall to the lot of any individual who would feel more compunction in proposing an alteration even upon an outward wall of a work executed under his directions. But such is the inconveniency and even danger to passengers attending the projection of the south-eastern angle of the parapet wall in front of the Register Office, that in the progress of these improvements the reporter would humbly propose, for the greater accommodation and comfort of the public, that this fine piece of masonry should undergo a small alteration, as represented in dotted lines upon the plan, in order to widen the street and improve the great thoroughfare to the port of Leith.

" Description of Line of Road recommended.

" In reference to the accompanying survey and plan, it will be proper to describe it more particularly. The first step towards forming the proposed new approach to the Calton Hill will be the removal of the houses which presently shut up the eastern extremity of Princes Street, and the other property in its direction eastward. The approach will then be made up to the proper level by a bridge extending in length about 362 feet from Shakespeare Square over Calton Street, towards the western extremity of the Calton Burying-ground, through which it will pass. Thence, passing in front of Bridewell, or between it and Nelson's Monument, it is continued along the southern side of the Calton Hill to the line of wall of division between the property of the city of Edinburgh and the lands of Heriot's Hospital. At this position the road begins to skirt along the southern side of the rising grounds in the parks of Heriot's Hospital, and crossing the eastern road to Leith it passes behind the houses of Abbeyhill, and ultimately joins the great road to London.

"The line of road just described has been laid out with gradients varying from 1 in 39 to 1 in 22. The more to the eastward the new line of road is carried before it joins the present London road, the more gradual and gentle the acclivity becomes. To improve this line of road still further by cutting deeper into the rock at the summit would not only create a great additional expense, but would place the road in a hollow, and shut out these characteristic views of the city which are the chief inducements to the new line of road.

" In determining the line of direction for the street from Shakespeare Square to Bridewell, it seems desirable that it should run in a straight line. The only objection to this is its interference with the Calton Burying-ground. In making any encroachments upon a place of burial, there is no doubt something very repugnant to the feelings, but in many cases this has been found necessary for public improvements, of which we have an example in the improved access from the bottom of Leith Walk to Bernard Street, where the road was carried through part of the churchyard of South Leith, and so in other parts of the country. The reporter has been at much pains in endeavouring to avoid the burying-ground, by attempting to turn the road more or less towards the left in going eastward, and by this means taking only a part from the northern side of that ground. But were the burying-ground to be encroached upon at all, and this cannot well be prevented, it seems less objectionable to carry the road in a straight line through it, especially as it may be found practicable to give an equal quantity of ground immediately contiguous to the present burying-ground without materially trenching upon any plan that may be in view for the erection of the prison; and as there will be a considerable depth of cutting in carrying the road through the burying-ground, the surface terring of the different places of interment may be removed to the new grounds with due care and becoming solemnity.

" The reporter gives a preference to this line, because it seems best suited to the peculiar situation of the ground, being calculated to show to much advantage the rugged rocks on which Nelson's

Monument is erected, which beautifully terminates the view in looking eastward; and in entering the town from the opposite directions, it exhibits at one view, from a somewhat elevated situation, the striking and extensive line of Princes Street. Now the reporter is humbly of opinion that to attain these objects, this line of road should be carried straight from Shakespeare Square to the eastern side of the burying-ground, after which it may be made to suit the position and nature of the ground in all its windings, as delineated upon the survey.

"As this road is not only to be the great approach from the eastward, but likewise to become the chief thoroughfare to the extensive lands of Heriot's and Trinity Hospitals, and to the lands of the other conterminous proprietors, henceforth likely to become the principal building grounds for this great city, which is always increasing towards its port of Leith, it becomes desirable for these purposes, and particularly to preserve the interesting view of the Calton Hill, that this road should not be less than seventy-five feet in breadth, or similar to Princes Street, exclusively of the sunk areas, which is certainly adequate to all the ordinary purposes of utility, intercourse, and elegance. There is, however, one way of viewing the width of this part of the road or street, by which it may appear to be too narrow even at seventy-five feet, and that is by comparing it with the width of Princes Street, which, including the sunken areas, measures ninety-five feet in breadth. Princes Street, however, comes more properly under the description of a row or terrace, and the principal footpath being on the north side of the street, it may consequently be apprehended that unless the new street were of an equal width, a spectator looking from the north side of the new street towards the line of Princes Street would command but an imperfect view of it. This to a considerable extent would be the state of the case even at seventy-five feet of breadth, and were the street reduced to sixty feet in breadth, as has been proposed, the view of the higher parts of the Calton Hill would be hid from the pavement on Princes Street. But the narrowing of the street even to sixty feet in width, with two elegant buildings in the form of pavilions or wings to the bridge, would have an effect similar to what is strikingly observable in looking from the western end of George Street towards the Excise Office. Examples of narrowing streets are not uncommon, as Great Pulteney Street in Bath, and Blackfriars and Westminster Bridge Streets in London. The reporter, however, confesses that he is not induced to consider sixty feet, or even seventy-five feet, as the most desirable breadth for the new bridge from any views of elegance ; with him the reduction of the width of the street is proposed rather from motives of economy to insure the success of a great measure, than from choice in making the design. In this situation a bridge of ninety-five feet, or equal to the extreme breadth of Princes Street, would most unfortunately place the new buildings upon the north-western side so near to the houses of Leith Street, that the windows of the houses of Leith Street and those of the new street would be shaded by each other, so as to require the buildings at the western end of the bridge to be kept less in height, if not to be discontinued altogether, for a considerable way, which would render the building grounds of much less value. Two or three of the new buildings, indeed, might be joined or connected with the old houses, but still the property upon the whole would be greatly injured. Considering this, and also the additional expense of the bridge without greatly increasing the value of the cellarage, together with the greater trespass that would be made on the burying-ground by a street of ninety-five feet in breadth, the reporter has been induced to delineate upon the plan a bridge of seventy-five feet, and a road from it to the Abbeyhill of sixty feet in breadth. Yet if it shall appear that funds cannot be conveniently obtained to meet even this expense, it may then be found necessary to make the whole of the uniform breadth of sixty feet. From the annexed estimate for the purchase of property, building a bridge of seventy-five feet in width, and making a road from it to the Abbeyhill of sixty feet in breadth, it appears that the expense will amount to £71,976, 14s.

"In estimating the expense of these works, the reporter has had in view that the road should be executed in aisler causeway, and that the whole should be executed in a substantial manner. From the borings in the strata which have been made by the directions of the reporter, there is reason to hope that the foundations of the bridge will not be difficult, and he therefore trusts that the several sums in the estimate of the expense already alluded to, will be found adequate to this purpose.

"The expenditure will no doubt be large, but the advantages are great in proportion.

"In considering this proposed new approach, it may be proper to notice it particularly as the means of procuring a proper site for the new jail and court house ; *second*, as calculated to raise the value of certain building grounds ; *thirdly*, as a public road ; and *lastly*, as contributing individually to the comfort of the inhabitants of Edinburgh.

"*Site for the Jail.*

"In any display of the advantages of this measure, the motive which led to it should not be overlooked. It was not the convenience of the wealthy citizen, nor the increased value of ground for building, nor even the improvement of the public roads that was sought after. It was to obtain a healthful situation for a *common jail*, and thereby to extend the comforts particularly of one unfortunate class of individuals, who, perhaps from the unavoidable circumstances of their lot, or from innocent misfortunes, are unable to pay their debts, and are cast into prison ; and even of another class, certainly less to be pitied, who from a perversity of disposition or the depravity of their nature, forfeit their liberty for a time.

" In looking for a proper site for building a jail upon the Calton Hill, the eye is naturally directed to the position of Bridewell as a fit place for concentrating the whole establishment of prisons for the city of Edinburgh to one spot, and if thought advisable, to put the whole under the care of the same governor, as is the general practice in England. A suitable site for the felons-jail has been pointed out upon the western side of Bridewell; and with a proper discrimination, the Sheriff proposes to erect the debtors-jail upon the other side; and if these buildings be constructed in the same style of architecture as Bridewell, the whole will present one uniform front or suite of buildings. The reporter understands, however, that the Sheriff does not wish this to be understood as fixed, but that the opinion of the most eminent architects should be obtained regarding the jail to be erected.

"Site for the Justiciary Court House.

" Supposing, for the present, that the jails were arranged in this manner, and that it were necessary in connection with them to erect a Justiciary Court House and public offices, a place must be found for them that shall at once be suitable in point of elegance, and be at the same time convenient for communicating with the prisons. In the event of adopting a street with a turn at the eastern end of the bridge, a site for these buildings could be very appropriately got, either facing the line of Princes Street or upon the southern side of the arch over Calton Street. On this last spot it may be objected that the buildings would not be fully seen till the spectator had reached the open arch of the bridge. Both of these situations would, however, be contiguous to the Register Office and North Bridge, and could be made accessible to the prisons by a private way round the southern side of the burying-ground.

" But certainly the most commanding site, in regard to elegance and grandeur of effect, for a public building would be to place it opposite to the prisons in the opening of the street, as marked on the plan. In such a position, when viewed from Princes Street in connection with the monument, the effect of these Court houses in perspective would indeed be very fine, and in coming round the hill by the line of road from the eastwards, it would be no less striking.

" The site for the prisons naturally points out itself contiguously to Bridewell, as well for the reasons already stated as on account of its southern exposure, and it has been observed to be just at the point of elevation for receiving a supply of water from the city's reservoir. But in setting down the public buildings for the county and for the Sheriff Court at so great a distance from the Court of Session and the other Courts of Law, the convenience of the practitioners is a consideration of importance which presents itself as requiring very mature deliberation, which does not strictly come under my notice.

" The value of Feuing Ground.

" The prolongation of the line of Princes Street by a bridge over Calton Street is calculated in a particular manner to benefit the extensive lands of Heriot's and Trinity Hospitals, and the conterminous proprietors to the eastward of the Calton Hill, by affording a better access than can be obtained in any other direction, especially in so far as it regards the higher grounds of Heriot's Hospital. But on this subject the reporter has already submitted his opinion in so far as regards Heriot's Hospital, in a report to the Governors of that institution ; and as the same argument held in a greater or less degree with the other proprietors, it seems unnecessary, in this place, to resume the subject.

" As a Public Road.

" As a new approach to the city of Edinburgh from the Abbey-hill to the central parts of the city, avoiding the inconvenient acclivity and awkward termination of Leith Street, or the still more intricate and incommodious access by the North Back of Canongate, this road will be regarded by the Trustees for the highways within the county as an improvement of the first importance. As a road, it is at once direct and obvious. By an extension of this line of road to Leith by the eastern road, or still more to the eastward through the lands of Restalrig, this access will be found of very general utility, while the traveller thus entering Edinburgh will be presented with the most characteristic views of the city, both old town and new town, calculated to inspire the highest opinions of its picturesque beauties.

" To the Inhabitants of Edinburgh.

" As a great addition to the individual comfort and convenience of the inhabitants of Edinburgh, the bridge over Calton Street will open an elegant access to the lands of the Calton Hill, from which the surrounding country forms one of the most delightful prospects of distant mountain ranges,—detached hills and extensive sea-coast, with numerous ships ever plying in all directions, together with the finest city scenery that is anywhere to be met with.

" Those who have admired the city of London from an eminence have indeed seen more extended lines of street bounded perhaps by a richer country, yet it is very deficient in that variety and boldness of feature which is so striking in this place. When it is wished to extend this walk to the eastward, the new road will lead the pedestrian commodiously to the bottom of Arthur's Seat, round the eastern side of which a path to Duddingston, branching out in various directions in its course round to Salisbury Crags, might, in a very delightful manner, be imagined to complete an afternoon's excursion. Let those who have not a lively picture in their mind of the prospect from the Calton Hill walk along the line of the projected road, and upon attending to it they will meet

with such a richness and variety of scenery as will satisfy them
how greatly the ornament of the city, and the pleasures of the
inhabitants and of its occasional visitants, would be promoted
by the continuation of the line of Princes Street towards the
lands of Calton Hill. Whether therefore we consider a bridge
over Calton Street as calculated to improve the approach to the
city from the eastward, or as rendering accessible many acres for
building, and villa grounds which must otherwise remain as grass
fields for an indefinite period, or as opening an easy way to the
rising grounds of the Calton Hill, in all these and in other important purposes the reporter is humbly of opinion that this measure
ought to be regarded as the greatest object which has engaged the
attention of public men since the erection of the North Bridge,
which was a very bold and enterprising undertaking for any period
of provincial or even of metropolitan history.

"Under these circumstances, it must be doubly gratifying to
learn, that notwithstanding the facility which an improved access
must afford in laying out the city grounds of the Calton Hill for
buildings, it is understood to be the intention of the Lord Provost
and Magistrates, in framing the Bill for an Act of Parliament for
regulating these works, to provide, with a proper liberality and a
due regard for the immediate and ultimate interests of the community, that these lands shall in all time coming be preserved open
and free as at present from all common buildings. It is also
hoped that the Hon. and Rev. Governors of Heriot's Hospital, with
enlightened sentiments, will preserve the view of Holyrood House
and its connecting scenery, by restricting the buildings on the
southern side of the new road through the Hospital's land to such
limits as may seem for that purpose to be necessary."

The Bill for this new approach to Edinburgh was
passed in 1814, and, on the 9th of September 1815, the
foundation stone of the Waterloo Bridge was laid with

great masonic ceremony, bearing the following inscription—

REGNANTE GEORGIO III. PATRE PATRIAE
URBIS PRAEFECTO ITERUM
JOANNE MARJORIBANKS DE LEES EQUITE BARONETTO
ARCHITECTO ROBERTO STEVENSON
CIVES EDINBURGENSES
NOVUM HUNC ET MAGNIFICUM
PER MONTEM VICINUM
AD SUMMAM URBEM ADITUM MOLITI
IN HOC PONTE NOMEN JUSSERUNT INSCRIBI
PROREGIS GEORGII AUGUSTI FREDERICI.[1]

which I quote, because Mr. Stevenson, in his notes, mentions a curious circumstance in connection with it :—
" The late James Gregory, then Professor of the Practice of Medicine in the University, the well-known author of the *Conspectus Medicinae Theoreticae*, was applied to by the Commission for the improvement to put the inscription in classical Latin. The Doctor came to me to say that he must style me *Architect*, there being no such word as *Engineer* to be found in the history of the Arts, and so it stands in the inscription. I wanted the Doctor to introduce the term Engineer, as it was very desirable to have the profession recognised in works now exclusively entrusted to the engineer."

[1] *Translation also by Dr. Gregory :*—" In the reign of George the Third, the father of his country, in the second year of the Provostship of Sir John Marjoribanks, Baronet, of Lees,—The citizens of Edinburgh having made this new and magnificent access over the neighbouring hill to the capital city, according to the plan of Robert Stevenson, Civil Engineer, ordered the name of the *Regent*, George Augustus Frederick, to be inscribed on this bridge."

M

Mr. Stevenson's original feuing plan, already referred to, for the Calton Hill had three ranges of terraces at different levels, as shown by a picture in my possession, from which Plate v. has been engraved. The middle line of terrace shown in the drawing corresponds to the Regent Terrace as ultimately constructed.

The approach on the northern side of the hill, known as the "London Road," was executed according to Mr. Stevenson's design immediately after the completion of the Regent Road and Waterloo Bridge ; and the whole of the new lines of road, as shown in red in Plate IV., were, as I have stated, part of the same design.

Mr. Stevenson's further contributions to the improvement of the approaches to Edinburgh were made between 1811 and 1817 to the "Trustees for the Post-road District of Roads," the "Trustees of the Middle District of Roads," the "Commissioners for forming and feuing Leith Walk," and the "Trustees of the Cramond District of Roads." These were the several authorities at that time in power, under whose directions he laid out the access to Edinburgh from Stockbridge by Royal Circus, and from Inverleith by Canonmills to Dundas Street, and from Canonmills to Bellevue Crescent. More recently the access from Granton Harbour to Inverleith Row on the east, and to Caroline Park on the west, were designed and executed under his direction in connection with his design for Granton Harbour, made to the Duke of Buccleuch in 1834.

To Mr. Stevenson's engineering skill, therefore, it may truly be said that modern Edinburgh owes much

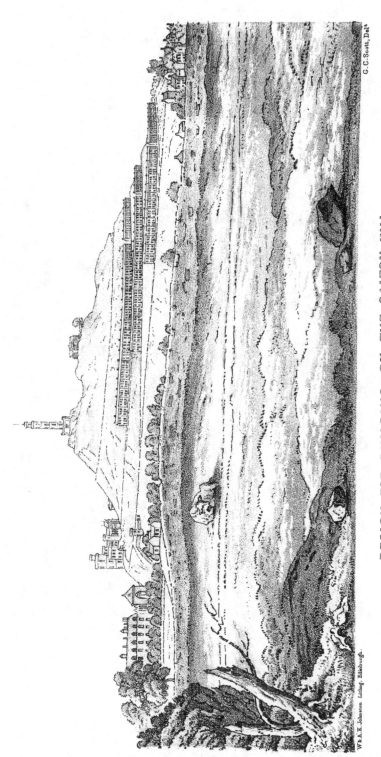

G. C. Scott, Del.ᵗ

DESIGN FOR BUILDING ON THE CALTON HILL.

by

Robert Stevenson, F.R.S.E. Civil Engineer

of its fame as a city of palaces, commanding views of the Firth of Forth and surrounding country which cannot be surpassed.

THE OLD TOLBOOTH PRISON.

While Mr. Stevenson was elaborating his designs for the new approaches to the city, his attention was naturally directed to the crowded state of the buildings in the old town ; and as we shall see, he did not fail fully to appreciate this evil, or forget to suggest a remedy for it in his plans of improvement.

The old "Tolbooth" prison, in the High Street of Edinburgh—the scene of so many incidents in the *Heart of Midlothian*—was still the only stronghold in which debtors and criminals were indiscriminately confined. Its position in the centre of the High Street, at St. Giles' Church, was very objectionable, and the erection of a new jail, in a more favourable situation, had been often proposed, but never carried out.

In pursuance of this desirable object, Sir William Rae—the Sheriff-Depute of Edinburgh—in 1813, accompanied by Mr. Stevenson as a professional adviser, visited many of the principal jails in England, including Newgate, Kingsbench, Cold Bath, Oxford, Gloucester, Chester, and Lancaster, to inquire into their general arrangements and accommodation.

Sir William Rae also remitted to Mr. Stevenson, in conjunction with Mr. Crichton, architect, to report on the condition of the ancient "Tolbooth ; " and from the conclusion arrived at by the engineer and architect, most

people of the present day will readily sympathise with
the Sheriff in his ardent desire for the erection of a
new building. Their report is curious, as conveying an
idea of the state of prison discipline in the early part
of this century, and is interesting in connection with
the antiquities of Edinburgh. Messrs. Stevenson and
Crichton say :—

"Agreeably to the directions of the Honourable the Commissioners for erecting a new jail, the reporters have examined
both the exterior walls and the interior parts of the present jail,
and they now report that this building, which was erected in
1562, originally formed the western extremity of a continuous
range of buildings in the middle of the High Street, called the
Luckenbooths. A few years ago these buildings were partly
removed, leaving the old jail in an insulated and unsupported
state. The street at the north-eastern angle of the buildings was at
the same time lowered several feet; and these changes, together
with the defective state of the masonry, appear to have produced
the following effects upon the eastern and northern walls of this
now shattered fabric.

"The eastern wall or gable is rent in three places. Two of
these fissures extend from the ground to the top of the building,
and the wall is found to bulge or bend outwards.

"On the northern side there has been a junction of the walls
of two separate buildings, forming what is called the *debtor* and
criminal sides of the prison, which seem to have been erected at
different periods. At this place there is a very apparent opening
from the bottom to the top of the prison, and the eastern or criminal end appears to be settling at the north-eastern angle, as further
appears from the doors of the guard-house and black hole, situate
in that quarter of the building, having at different times required
some alterations to make them move upon their hinges.

" This wall, like the eastern one, is also bulged outwards to the extent of from six to ten inches in different places.

" The roof of the prison is likewise unsafe, particularly upon the criminal or eastern side, where the rafters have sunk in the middle and pressed the side wall outwards at the top.

" *Interior of the Prison.*

" Upon examining the interior of the prison, it was found that the several *cracks* and *fissures,* already described as observable on the outside of the building, were also most distinctly visible from within, and that the northern wall in several of the apartments appears to have separated from the floors. In confirmation of these facts, which appear particularly to claim the notice of the Honourable the Commissioners, it was distinctly stated to the reporters by Mr. Sibbald, the head jailer, that he had been conversant with this prison about twenty-two years; that about seven years ago he became principal jailer, and had ever since been in the habit of making requisitions for the necessary repairs, which were always executed at his sight; that these openings and fissures, which now appeared obvious to the reporters, had been frequently plastered over with lime, sometimes previously to whitewashing the apartments, and at other times at the earnest request of the prisoners, to stop the current of air, which annoyed them, and still these fissures appeared to be getting wider; that in every instance where the walls had been attempted to be forced by the prisoners, the mortar was found to be loose and soft, without having taken bond; in particular it was stated that two of the prisoners had lately excavated about two cartloads of rubbish from the walls with a small piece of iron, in the course of a few hours.

" It is therefore humbly concluded, from the information obtained by the reporters upon the spot, but especially from their own knowledge and observation, that there are data for assuming that the eastern and northern walls of the prison have deviated considerably from the perpendicular of their original elevation; that there is reason to consider them still continuing to deviate from the perpendicular; and that finally, in the decayed state of this building, it is impossible to warrant its stability for any given period of time. The reporters should even consider the continuing the use of this building for one year longer than is indispensably necessary for the erection of a sufficient jail, an evil if possible to be avoided, as involving imminent danger to the wretched inmates, and much hazard to the public at large, from its position in the heart of the city.

"It would accordingly be very desirable that some support could be given to this old building immediately, but unfortunately its position renders this quite impossible without seriously obstructing the High or principal street of the city. Had it formed any part of the Honourable the Commissioners' instructions to the reporters to take notice of this jail as a place of security for the safe custody of prisoners, it would only be necessary for them to refer to what is herein stated regarding the insufficiency of the walls, and to remark that the floors, being wholly composed of timber, are neither proof against the simplest accident by fire nor against the slightest attempt at escape by the prisoners. It is truly surprising that any criminal of a desperate character can be retained within its precincts to abide the pains of law, which nothing but the active vigilance of its keepers could insure."

This ancient prison-house was removed in 1817, and in his Notes to the *Heart of Midlothian*, Scott says :— "That with the liberal acquiescence of the persons who had contracted for the work, he procured the stones which composed the gateway, together with the door and its ponderous fastenings, to decorate the entrance to the kitchen court at Abbotsford."

THE REMOVAL OF THE COLLEGE.

There is yet another report which, though its interest may only be local, I think is worthy of a place in this Memoir, as it not only shows Mr. Stevenson's firm conviction in the ultimate success of his Calton Hill improvements, but is a pleasing record of his interest in the scene of his early studies.

It is not, I believe, generally known that Mr. Stevenson made an unsuccessful attempt to have the University buildings, then in progress, removed from the old town to the site proposed to be opened up on the Calton Hill; and the remarks he then made, addressed to the Right Honourable Sir John Marjoribanks, Lord Provost of Edinburgh, may have interest even at the present day, as shadowing forth views which, in the now altered relations of the new and old town, have been to some extent realised.

"In making the following observations at the desire of the Lord Provost regarding the completion of the College of Edinburgh, the memorialist would be understood as referring to the *site* of the building rather than to the merits of any particular design, of which he does not presume to give any opinion, as it is a matter which more properly falls under the observations of the architect than the engineer.

"In treating of the fitness of the present site of the College of Edinburgh, it may be proper to take some cursory notice of the situation of the Old College, as connected with the houses and streets in the neighbourhood, and then show the alterations which the University grounds have undergone since the design was first formed of rebuilding the College.

" *Old College.*

"In so far as the memorialist can recollect the exterior of the area of the Old College, it was occupied by a range of *low* buildings of only two stories, particularly upon the southern and western sides, and was again divided by a range of buildings into a small lower court towards the north, and the present main court-yard on the south, and these two courts communicated with each other by a spacious flight of steps, so that the principal or higher court was comparatively open and free to the influence both of the sun and of the air. Nor was there any obstruction to this state of things beyond the precincts of the College for a consider-able period after the New College was commenced, and until the elegance of the building stamped a new value upon all the sur-rounding property. But, unfortunately, by this time the funds for the works fell short, and the operations were stopped. The Magistracy, also, who originally entered upon this great work, in rotation retired from office, and the same zeal was perhaps not felt by those who immediately succeeded; and we are now left to regret the shortness of the period of human life, which has removed the man who conceived the magnificent design of this building, which is now so completely invested with streets as to be rendered nearly unfit for the purposes of its foundation.

" *New College.*

"The *site* of the New College of Edinburgh, as already stated, does not possess any of those properties which are considered essen-tial to the convenience and eligibility of a public school. Instead of being in a retired situation with sequestered walks, like the other colleges of the United Kingdom, it is closely surrounded by paved streets, which are the most public thoroughfares for carriages in the city, insomuch that the memorialist has witnessed the annoy-ance of Playfair's mathematical class by a ballad-singer, and he has oftener than once seen the Professor of Moral Philosophy put to silence by the disloading of a cart with *bars of iron* in College Wynd; and at all times the driving of a single carriage briskly in the streets which surround the College is sufficient to disturb, and even to interrupt, the classes. To this it may be replied that double windows will prevent such interruptions; but these would obscure the light which already, from the late erection (on all sides) of very high buildings, is much injured.

"So strongly is the memorialist impressed with these views, from what he has himself as a student experienced, and from what he has heard from others, that he cannot resist bringing them forcibly under the notice of your Lordship in connection with the erection of a building for one of the first seminaries of education in Europe.

"When your Lordship's predecessors in the office of the magistracy adopted the plan of Robert Adam, the most eminent and justly celebrated architect of his day, the site was comparatively free from the objections stated. It is not therefore the plan which is objectionable, but it is the neighbourhood which has been so altered and changed as to be very unsuitable to the elegant design of the architect.

"From causes to which it is unnecessary to allude, the building of the New College has only advanced about one third towards the perfecting of the design, and a sum of money is now expected to be procured for its completion. The present moment is therefore one of the greatest importance for considering the deficiencies of the present site, and if found materially defective, as humbly appears to your memorialist to be the case, it were much better to change the site of the building while it may be done without much loss, and execute the design in a more eligible situation.

"It must always be kept in view that when this design was made the grounds were open to the free circulation of the air and the full influence of light. But now the case is materially altered,

N

and if the design is executed under such a change of circumstances the direct rays of the sun will hardly ever reach the area of the courtyard, especially in the winter months, neither will there be that free circulation of air which is essential to health and comfort, and moss (byssus) will make its appearance upon the lower parts in the interior of the courtyard, which is very unsuitable in a magnificent building such as Mr. Adam's design for the College of Edinburgh.

"At the period when the rebuilding of the College was determined upon there was perhaps little choice as to the spot for its erection; the number of students, now greatly on the increase, was at that time much smaller, and the College grounds were then much more relieved and uncumbered with other buildings, a state of things which most unquestionably would have been preserved had the building proceeded as was expected; but in the lapse of about one third of a century many changes take place, and the slow progress of the building necessarily produced a want of energy in the official people to prevent the use that has since been made by the respective proprietors of the surrounding grounds.

"At the present crisis, however, your Lordship will now feel yourself called upon in a review of these circumstances to consider what is proper to be done upon a great scale for the ultimate best advantage of future generations in a matter of great public interest. Under these impressions a field of operation is just opening for your Lordship's consideration, in a prolongation of Princes Street in a direct line to the lands of Calton Hill and Heriot's Hospital, now in progress under the auspices of your Lordship. To take a minute view of this improvement would be tedious, and would require the notice of more particulars than these observations are intended to refer to. But in a general way it may be noticed that there is ample space and freedom for the execution of Mr. Adam's design on the lands to which the new approach will lead by a very easy access.

" It may be objected to the removal of the College that it would be inconvenient for the students ; but for those who are perhaps the most numerous, living in the New Town, a site for the College on the north side of the town would be the most convenient, and for a different class lodgings at a cheap rate would be procured quite at hand in the Canongate.

"A more powerful objection would perhaps arise from the contiguity of the present site of the College to the Infirmary and other institutions connected with the education of the medical classes, but these may also be got over by a little arrangement in the present hours of the classes, and one would not despair of seeing a more direct road projected from the Calton Hill to the southern side of the town were the College removed to that neighbourhood. With regard to any real loss to the students, it is not believed that such could be instructed were this proposition fully considered. But those who would perhaps be the most clamorous are the persons who have made the most of their property by building immense piles of lodging-houses in the immediate vicinity of the College, and have thus ruined the neighbourhood.

"With regard to the funds for this change of site, your memorialist is of opinion that the removal of the College from the present valuable grounds in the central parts of the city, for buildings applicable to commercial and economical purposes, would be attended with an increase of funds towards the new erection ;—for the lower part all round would be opened for valuable shops, while the higher parts would answer for dwelling-houses and other purposes. The part of the front would be easily convertible into a house for the Royal Bank, which seems much wanted, and in short it may be confidently stated that upon the whole there would be no loss, but gain, by the change of position, while very many advantages could be pointed out as attending such a measure, were this the proper place for entering more fully into the subject.

"The proposal stated is not new; it has been often under the memorialist's consideration, and he has heard it favourably spoken of and received by several of the Professors of the University, in particular Professors Leslie and Playfair, and others eminently qualified to judge correctly upon the subject."

With this report I conclude what may be fairly held to be of purely *local* interest, but which nevertheless I have thought worthy of a place in the memoir of one whose great anxiety ever was to secure the amenity of Edinburgh, and make it attractive not only as a place of residence but as a seat of learning.

CHAPTER VI.

FERRIES.

Ferry Engineering—Extracts from Report on the Tay Ferries—Reports on
various Ferries—Orkney and Shetland Ferry, etc.

BEFORE we had steamers to navigate our firths and
railways to bridge our estuaries, the "crossing of the
ferry" was an event of no small solicitude to the traveller.
In the sailing pinnace-boat of those days he not only
might encounter serious danger, but his exposure to sea-
sickness and drenching spray depended wholly on the
weather, and sometimes the length of the passage, and
the duration of his suffering could not be foretold by the
most experienced "Skipper," as the captain of the boat
was invariably styled. Anything that could reduce the
hazard and uncertainty of so miserable a state of things
was naturally hailed as a priceless boon; and the improvement of "ferry communication" at the beginning of this
century was an important branch of civil engineering.
Its successful practice demanded nautical knowledge as
well as constructive experience, for the engineer had first
of all to study the strength and direction of the tidal
currents of flood and ebb, and then to consider from what
points on the shore a ferry-boat, under the varying
states of wind and tide, could most readily make her
passage across. He had further to select the most suit-

able sites for landing-places, and to construct high and low water *slips* at different points to meet the varying states of tide and wind, and to construct roads of more or less extent to connect the landing-places with existing turnpikes. All this arrangement was required, because at the time of which I write, before steamboats were invented, two costly deep-water piers placed *ex adverso* of each other, one on each side of a ferry, would not have met the requirements of the case; for the management of a sailing pinnace, at the mercy of the currents and winds, demanded not a single pier for which to steer, but a choice of several points, on as wide a range of coast as possible, for which the "skipper" could shape his course and make a landing. Mr. Stevenson's nautical experience peculiarly fitted him for giving valuable advice in this important branch of marine engineering. It is no doubt a branch of the profession which may be said to be obsolete, but I do not know that on that account it is undeserving of notice; and the best mode I can think of for conveying to any one who may be interested in it an idea of the "ferry engineering" of former times, is to give an extract, with an illustrative sketch, of one of Mr. Stevenson's early Ferry Reports. I select for this purpose a report made to the "Freeholders, Justices of the Peace, and Commissioners of Supply of the counties of Fife and Forfar" relative to the ferries across the Tay at Dundee :—

"Having examined the shores and firth of Tay the reporter has now the honour of submitting the following as his report regarding the proposed improvements :—

"The improvement of the ferries on the Tay has long been the desire of the public; and though this measure has hitherto been delayed, on account of the expense which necessarily attends such operations, yet so desirable an object has been invariably kept in view; and now, when the advantages attending the recently improved state of Queensferry and Kinghorn ferries have been in a good measure realised, the passage across the Tay has very opportunely been brought under the consideration of the freeholders of the adjoining counties.

"The present landing-slips or quays upon the Tay are situate at Dundee upon the north, and at Woodhaven and Newport on the south. The bed of the firth or river at Dundee is so much silted up and encumbered with sandbanks and mud, that the piers, which were no doubt originally built of sufficient extent, and perhaps commanding the necessary depth of water for floating the passage-boats at low tides, have at length become inadequate to so great a thoroughfare, and the boats are now left by the water at every spring-tide, to the great annoyance and inconvenience of the public.

"It will be observed from the plans accompanying this report that the Craig pier at Dundee is proposed to be extended from the southern extremity of the present landing-slip or pier 400 feet in length, or to the southern extremity of the Craig rock, so as to command a depth of about five feet at low water of spring-tides, which will be sufficient to float decked boats of twenty to twenty-five tons register, built upon a suitable construction for sailing. It is proposed to construct this pier, where the greatest business is to be done, upon the plan of a double pier, sixty feet in breadth; and as it will now be of a much greater extent than formerly, a *screen wall* is proposed to be erected in the middle of it, in a longitudinal direction, so as to check the waves or run of the water over the pier, and also for the defence and shelter of passengers from the inclemency of the weather. This pier will form an inclined plane sloping to seaward at the rate of one perpendicular to twenty-six horizontal.

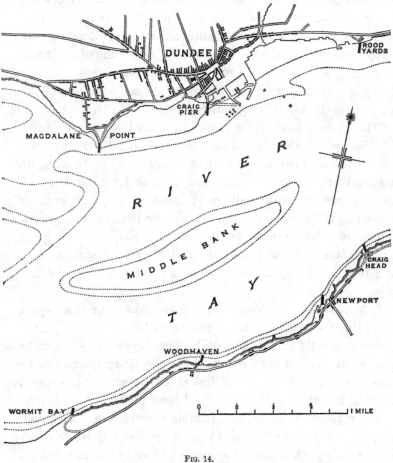

Fig. 14.

"In sailing from the southern side of the Tay for Dundee, it will on some occasions be found convenient, with certain directions of the wind and currents of the tide, to have landing-slips or piers both above and below the town of Dundee, so as to prevent the necessity of *tacking* with adverse winds, as is the case at present from the want of such accommodation. Upon examining the shores above or to the westward of Dundee, the most convenient position for a landing-slip is at the Magdalene Point, about 1400 yards to the westward of the Craig pier at Dundee.

"In the same manner a convenient position presents itself on the rocky shores of the Rood Yards, about 2000 yards below or to the eastward of the Craig pier. These proposed new piers are delineated and laid down in the drawings accompanying this report.[1]

"Upon the southern side of the Tay, and opposite to Dundee, the harbour or landing-slip most frequented at present is that of Newport. In former times, when the accumulation of sand, called the Middle Bank, between the opposite shores of the ferry at Dundee, was less extensive, the principal landing-place upon the Fife side of the firth was that of Woodhaven. Newport is a small harbour, built of masonry, with a landing-slip or sloping pier attached to the outward wall of the harbour for the convenience of the ferry-boats. At this station it will therefore only be necessary to extend the landing-slip about eighty-eight feet northward, in order to obtain five feet of depth at low water of spring tides; and as the present sloping pier or slip is inconveniently narrow, it is proposed to add fourteen feet to its breadth; and the reporter would recommend that this work, in connection with the pier at Dundee, should be executed in the first instance, on account of its being of primary importance in the improvement of the Ferry.

"At Woodhaven it is proposed to add seventy-one feet to the length of the landing-slip, to enable the ferry-boats to approach it at low water of spring tides, in the same manner as at Newport.

"At or near Craighead, about 830 yards below or to the eastward of Newport, there is a convenient point of land, where it is proposed to erect a slip or pier 250 feet in length and 30 in

[1] From which Fig. 14 has been made.

O

breadth. This pier will command five feet, or a sufficient depth of water for the ferry-boats at the lowest tides, and is in a position calculated to be highly useful.

"A pier has likewise been suggested as necessary at Wormit Bay, about a mile to the westward of Woodhaven, which, in certain directions of the wind, may no doubt be found useful; but when the piers opposite to Dundee come to be put in good order, and the ferry placed under proper regulations, it is presumed that a pier at Wormit Bay would very seldom be found necessary. The cost of these works is estimated at £20,952, 13s. 6d.

"In forming the several landing-places already described, it is obvious that there must be a ready communication between each of these piers and the public roads in their respective neighbourhoods. It will also be of essential importance to this measure, that a connection by good roads be formed and kept up between the several landing-places, in so far as this can be effected. At present there is a pretty good line of road between Woodhaven and Newport, which would require to be extended eastward to the landing-place at Craighead.

"In the event of Craig pier being adopted as the landing-place at Dundee, it might be advisable to take a power in the proposed Act, as a measure of the burgh of Dundee, for making a new and more direct approach from that pier to the main street. The extension and formation of these roads, however, will necessarily fall under the joint consideration of the trustees for the ferries and roads in apportioning the expense between the respective trusts.

"*Boats.*

"At present there are said to be no less than about thirty boats plying upon the passage at Dundee, which are navigated by about fifty men and boys. But were the piers and landing-places, with the accesses to them, completed in the manner proposed, and the whole placed under proper regulations, there can be little doubt that the ferry of Dundee would be much better attended, and the public better served, by one half of the present number of boats, as has been experienced on the ferries of the Firth of Forth.

"*Steamboats.*

"Some are of opinion that both the number of boats and of piers or landing-places might be still further reduced by the introduction of the *Steamboat* upon this passage. The reporter, however, does not think it would be advisable to have fewer than three landing-places at each station, as even the steamboat itself is more or less liable to fall short or to be driven past its port by adverse winds and strong currents; and, in a great public measure of this kind, it is proper to be prepared for the worst that is likely to happen. Regarding the adoption of the steamboat in preference to sailing-boats, the reporter is not however prepared to give any very decided opinion upon the subject. He has, indeed, seen the steamboat used with great facility on the passage across the river Mersey at Liverpool, and has himself brought the plan of a steamboat under the notice of several of the trustees for King-horn and Queensferry passages, proposed to be constructed upon similar principles with that originally tried, it is believed, by the late Mr. Millar of Dalswinton. But it would seem to be premature to recommend the framing of the Bill or the construction of piers for Dundee ferry upon the idea of the exclusive use of the steam-boat. The consideration of the late unpleasant accidents which have befallen some of those boats renders this a matter of great delicacy, and one in which much precaution should be used on so public a ferry. Under such circumstances it is not only necessary to consult the actual safety of passengers while afloat, but even to meet their prejudices, with proper attention to their comfort. From considerations of this kind, the reporter recommends that such of the piers or landing-slips on the ferry of Dundee as may ultimately be erected, should be completed agreeably to the plan herein proposed; and it is fortunate that, with some trifling alterations or additions, the piers suitable for the common boat can be made answerable for the steamboat. When this measure is in full operation it may then be highly proper to make an experiment with the steamboat upon the passage at Dundee, and if this mode is approven of by the public it can be extended, and the number of sailing-boats diminished accordingly."

Mr. Stevenson was employed to give similar advice by other Trusts, and particularly by the "Trustees of the Queensferry Passage" and the "Trustees of the Edinburgh and Fife Ferry," both across the Forth,—the "Freeholders and Justices of Peace of the counties of Ross and Sutherland," for the Ferry of the Dornoch,—the "Freeholders of the county of Glamorganshire," for the new passage-ferry of the Severn, to all of whom he made reports at various times, as to the improvement of the mode of communication under their charge. He also was engaged by the Lords of the Treasury "to inquire into and report on the best mode of improving the post-packet communication to Orkney and Shetland," which he did after careful survey and consideration, in an elaborate report, from which I give the following extracts, as illustrating some of the disadvantages under which the public laboured before steam was generally adopted :—

"The islands of Orkney are separated from the coast of Caithness or mainland of Scotland by the rapid channel of the Pentland Firth, which varies in breadth from six to nine miles, while Zetland lies fifty miles to the northward of Orkney."

"These two groups of islands, forming one county, are of late years greatly advanced in importance, and possess an aggregate population of 60,000 inhabitants, who are chiefly engaged in maritime affairs and fishing adventures. From their local position also in the North Sea, they lie much in the track of vessels sailing in the higher latitudes, and correspondence with them regarding the destination and insurance of ships is often of the greatest importance to commercial men. It is likewise known to the Right Honourable the Lord Advocate of Scotland, and the Honourable the Sheriff of the county, that the want of a proper communication by post not unfrequently interferes with the regular administration of justice in these islands; and now that Orkney and Shetland jointly send a member to Parliament, the evils resulting from the want of a regular communication press more forcibly, not only on the inhabitants of these islands, but on the public generally.

"So uncertain is the post of Zetland on its present footing, that the reporter himself carried to Lerwick the first intelligence of the appointment of Sir William Rae as Lord Advocate of Scotland, after it had been currently known through the newspapers in all other parts of the kingdom for several weeks, and it is well known that the succession of the King was not known in Lerwick for several months after the event took place. During the winter months the intercourse is indeed precarious as well as uncertain, and much painful delay is often experienced by parties interested in any question connected with the insurance of vessels wrecked on this dangerous coast.

"In order to lessen the labour and expense to themselves, the Orkney ferrymen on either side contrive to leave their shores so as to meet about the middle of the Firth, where they exchange the mail and passengers, and then return to their respective homes. In this way they seldom complete the full trip across the Firth, excepting when obliged by stress of weather. This interchange of the post from boats, it must be allowed, is rather a hazardous experiment anywhere, but more especially in the middle of the Pentland Firth; and whether the inhospitable state of the shores on either side, the rough and boisterous nature of the sea to be passed through, or the want of management be considered, there is evidently great room for improvement on the ferry of the Pentland Firth."

This communication is now, as is well known, carried on by first-class steamers, which touch at Kirkwall and Lerwick, and by a daily mail steamer which crosses the Pentland Firth from the low-water pier at Scrabster in Caithness to Stromness in Orkney; and the travelling public may be congratulated that the ferry communication of the early part of the century, of which I have given a sketch in this chapter, no longer forms a part of the practice of the civil engineer.

CHAPTER VII.

RAILWAYS.

1812—1826.

Canals and Railways on one level—Haulage on Railways—Railways in Scotland
—Edinburgh and Midlothian, Stockton and Darlington, and Edinburgh
and London Railways—Uniform gauge proposed—Notes on Railways for the
Highland and Agricultural Society—Letter from George Stephenson.

GREAT powers of observation, combined with fertile and
practical mechanical resources, enabled Mr. Stevenson in
many cases to form engineering opinions which may truly
be said to have been " before their time," and in no subject, perhaps, was this more strikingly realised than in his
views as to railways.

Impressed with the great inconvenience of change of
level in canals, involving "lockage," with all its expensive
works and serious obstruction of traffic, he early formed a
firm belief that wherever lockage could be avoided, by
making even a considerable detour in the line of canal, it
was sound engineering to adopt the level line, although it
might be at the cost of additional length. Founding on
this general opinion, so early as 1812, he traced out and
proposed lines of canal to be carried upon *one level*,
without lockage, through the valleys of Strathmore and
Strathearn, connecting Perth, Forfar, Arbroath, and
Montrose, and also by a line of canal, by Broxburn,

Linlithgow, Polmont, Castlecary, Campsie, and Broomie-
law, to unite Edinburgh and Glasgow.

His early researches on the subject of canals prepared
him, about 1816, to extend the same reasoning to railways,
which, with wonderful sagacity, he foresaw must become
what he termed the "British highway" of the future.
He found that his first idea of tracks of iron and stone to
improve the draught on common roads was not destined
to meet the requirements of the future ; and when as yet
nothing was known of railways beyond the tramways connected with coal-fields, and no proposal had been made
to adapt them to passenger traffic, Mr. Stevenson was
engaged tracing in all directions through Scotland lines of
railway as a new mode of conveyance to supersede roads.
Some of these early proposals, extending to about five
hundred miles, are shown in hard lines on Fig. 15, and
of all these railways he made surveys, estimates, and
elaborate reports addressed to Committees of subscribers
by whom the various schemes were supported.

It must be remembered that at that early period no
other power than that of horses was contemplated for
performing the haulage either on road, canal, or tramway,
and Mr. Stevenson, true to his early views as to the disadvantage of lockage on canals, spent much time. in ex-
perimenting on the prejudicial effect of steep inclines on
horse railways, and in endeavouring, in his various surveys,
to discover routes by which his lines of railway might be
carried through, as much as possible, on one level, regarding a few miles additional length of line as quite unim-
portant compared to the disadvantage of a steep gradient,

—a view which was more appreciated before the locomotive engine had taken upon itself the labour of the horse.

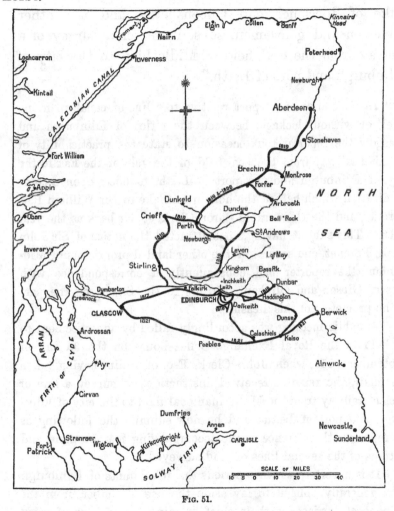

FIG. 51.

To show the state of railway matters at the period to which I refer, I think it may not be uninteresting to give,

P

even at some length, extracts from Mr. Stevenson's report on what was called the Edinburgh Railway. The report, which is dated 1818, was addressed to "His Grace the Duke of Buccleuch and Queensberry, and the other noblemen and gentlemen, subscribers for a survey of a railway from the coal field of Midlothian to the city of Edinburgh and port of Leith."

"In the course of a report relative to a line of canal upon one level, or without lockage, between the cities of Edinburgh and Glasgow, the reporter took occasion to state the practicability of a line of railway from the coal field of the vale of the Esk to the city of Edinburgh and the port of Leith, founded upon a communication which he had the honour to make to Sir William Rae, Baronet, and the Honourable Baron Clerk, so far back as the year 1812. This subject having since attracted the notice of Sir John Hope, Baronet, and several of the other landed proprietors of Midlothian, the reporter had consequently a correspondence with Messrs. Gibson and Oliphant, Writers to the Signet, on the part of the promoters of this measure.

"A public meeting was accordingly called by advertisement to be held in the Royal Exchange Coffee-house on the 3d day of September 1817, when John Clerk, Esq. of Eldin, having taken the chair, the reporter received instructions to survey a line or lines of railway from the Midlothian coal field to the city of Edinburgh and port of Leith; and he now submits the following as his report, with reference to the accompanying map or plan, and sections of the several lines of road surveyed.

"It is uncertain at what periods the inhabitants of Edinburgh were generally obliged to lay aside the use of timber, from the distance of carriage, as their chief building material, or of wood and turf as fuel; neither have we any certain information at what time pit coal was discovered, or the coal field of the Lothians first opened. But it is in the recollection of some persons still living, that, owing to the miserable and circumscribed state of the roads, or rather the want of formed roads altogether, pit coal continued to be conveyed in sacks and on horseback for supplying the city of Edinburgh. These horse tracks, originally taken up by accident, were persevered in by obstinate habit; and being afterwards followed as the lines of our future roads, have become the ultimate source of much of the difficulty attending their improvement, from the soft and miry track of the pack-horse and the sledge, to the broad and spacious *stoned* carriage-way, in combination with the trim footpath of the present day. But, even here, experience shows that it would be improper to rest satisfied, and cease from further exertion. The acclivities of the road may still be levelled, and its asperities smoothed, by the introduction of the more compact and durable materials of the *British Roadway* or Iron Rail. Such, however, has been the progressive nature of discovery in all ages, that we are only beginning to appreciate the immense advantages which would attend the introduction of a new system of roads or railways, laid upon a level or horizontal base, as admirably calculated to increase the power of the horse in a tenfold proportion by destroying friction—that bane to animal labour as now applied on the common road.

"Wagon-ways constructed entirely of square wooden frames or rails, laid in two right lines on wooden sleepers, appear to have been in use at Newcastle so far back as the year 1671. The plan of cast-iron railways seems to have been originally introduced by the great Iron Company of Colebroke Dale in Shropshire, only about the year 1786, as an improvement upon the tram or wooden railway; and such are likely to be the benefits resulting from this discovery, that we doubt not, as this system develops itself, the name of the person who first conceived the idea will eagerly be sought after, and honour done to him, as to one of the greatest benefactors of his country. We might mention the name of the late Mr. Jessop, as the first engineer of eminence who seems to have introduced railways in the south. He was also the engineer for the magnificent works of his Grace the Duke of Portland in Scotland, connected with which there is a double railway from Kilmarnock to Troon, which is ten miles in length. The other railways in Scotland of any extent are those at the works of the Carron Company, Lord Elgin's, Mr. Erskine of Mar's, Sir John Hope's, and other coal works. A public railway has also been projected from Berwick-upon-Tweed to Glasgow, an extent of country of about 125 miles; and an Act of Parliament has already been obtained for completing part of this track, viz., from Berwick to Kelso.

"A railway has the advantage of being formed at an average of one third perhaps of the expense of a navigable canal; and in many situations its first cost may even be compared with the expense of making a common road. The result is also favourable if we inquire into the comparative quantities of work done upon a canal and a level railway. Upon the canals in England, a boat of thirty tons burden is generally tracked by one horse, and navigated by two men and a boy. On a level railway, it may be concluded that a good horse managed by a man or lad will work with eight tons. At this rate the work performed on the railway by one man and a horse is more than in the proportion of one third of the work done upon the canal by three persons and a horse, if we take into account the more speedy rate of travelling and the facilities to general trade in loading and discharging, together with the difference of the first cost of a railway, which altogether give it in some cases a decided advantage over the navigable canal. If we compare the railway with the common road, it may be fairly stated that, in the instance of a level railway, the work will be increased in an eight or ten fold proportion. The best horse, indeed, with difficulty, works with three fourths of a ton on the common road, from the undulating line of its draught, but on a level railway it is calculated that he will work even with ten tons. But to increase the economy of the railway system still further, we have only to employ one man to work two horses.

"*Line of Draught.*

"With regard to the line of draught, or longitudinal section of a railway, it may be stated as one of its great advantages that it is more easily accommodated to the irregularities of the ground through which it has to pass than a navigable canal; and even where the ground is so irregular as not to admit of a uniformly level track, or an inclined plane, there are several simple methods which may be resorted to for lifting the wagons from one level to another, so as to produce similar effects with lockage on a canal. In so far, however, as the present design of the Edinburgh Railway has been carried by actual survey, neither of these plans will be found necessary upon the main lines. Even on the descending line, the fall is so extremely gentle that the horses in returning may be loaded with four or five tons. But the proposed mode of lockage may with propriety be introduced on the several offset branches, such as those from Leith to the main line, and from Monkton Hall and the Cowpits to Dalkeith, and to the southern parts of the county, on which a trade may be expected to be carried both to and from the main line.

"Where the load or trade is all in one direction, it is a maxim in *practice*, that the fall should be so apportioned to the rise, that the work may be equal *down* with the load, and *up* with the empty wagons. But where there is to be a trade both ways, it is obviously much to be desired that a level in all such cases should be obtained. This, in the Edinburgh Railway, has been found from the declining aspect of the country towards the sea; but as there will be less return trade on this railway in merchandise and manure, etc., to the eastward, than the coal and building materials, etc., carried to the city, it becomes a question of policy how far it may be proper, in this instance, to adopt the level line at a great additional expense.

"By the level line to Edinburgh the branch to Leith becomes also somewhat more lengthened than by the descending line, which, instead of preserving the level, is always falling, or approaching towards Leith. The reporter, as before noticed, has various modes in view, by which the branch to Leith may be made of a very easy line of draught, or be thrown into a succession of levels, by a species of lockage or stepping. Where sudden acclivities occur on the line of a railway they are generally overcome by an inclined plane, of greater or less extent, according to the particular rise, and on this the loaded wagons are brought up by a steam-engine. But to render railways applicable to all situations, it seems to be necessary that the overcoming of such obstacles should be within the reach or power of the driver and his horse; by working a kind of *gin* connected with an inclined plane, or by lifting the loaded wagons perpendicularly, which may in various ways be accomplished by the aid of pulleys, by the common lever, or the revolution of a wheel.

"This subject has been justly considered to be a matter of so much public importance, that the Highland Society of Scotland has offered a premium for an Essay, with models, for lockage on railways; and the reporter has no doubt that by this means much additional light will be thrown on the subject.

"There are few subjects on which those conversant in the working of draught animals are more divided than about the proper *line of draught*. Some do not hesitate to affirm, that a level road is injurious to the horse, and that an undulating road is preferable to one by which the ascent is long, though gradual. Such are of opinion, that by throwing the road into successive eminences, or *up and down hill*, various muscles are brought into action, while others are left at rest, and this alternation they conceive to be the best condition of things for the animal.

"Being rather, however, at a loss in regard to that part of the subject which relates to the operation of the muscles, the reporter applied for a solution of the case to a distinguished medical friend in this city [Dr. John Barclay], eminent for his knowledge and for his great exertions in the science of *Comparative Anatomy.* His answer to the queries which he allowed the reporter to put contain the following comprehensive passages : ' My acquaintance with the muscles by no means enables me to explain how a horse should be more fatigued by travelling on a road uniformly level than by travelling over a like space upon a road that crosses heights and hollows ; and it is demonstrably a false idea that one set of muscles can alternately rest and come into action in cases of that kind. The daily practice of ascending heights, it has been said, gives an animal *wind,* and enlarges the chest; it may also with equal truth be affirmed that many horses lose their wind under this sort of training, and irrecoverably suffer from imprudent attempts to induce such a habit.' In short, he ascribes much to prejudice, ' originating with the man, who is continually in quest of variety, rather than the horse, who, consulting only his own ease, seems quite unconscious of Hogarth's *line of beauty.*'

"In the course of investigating the subject of the draught of horses, the reporter has made several experiments with the dynamometer, both upon canals and railways, with a view to ascertain the power of horses and the best line of draught; and he has further the satisfaction to find, that the result of these trials agrees nearly with experiments made, and obligingly communicated to him, from various parts of the kingdom. The reporter therefore concludes that the force with which a horse will continue to work is about one-sixth or one-seventh of his absolute weight. Now, as he found the average weight of three ordinary cart horses to be about ten cwt. it may be assumed, generally, that a horse can continue to work with a force equal to 160 lb.; and allowing 40 lb., or one fourth, for friction, there remains 120 lb. to be applied to the load. In these trials, when the wagons were put in motion, it appeared, under favourable circumstances, that a force of about 12 lbs. only was necessary to move one ton upon a level edge railway, which by calculation would give about ten tons as the load of a good horse weighing ten cwt. ; but, for practice, this will perhaps more properly be taken at about eight tons. With regard to inclined planes, it may be noticed, that for every one fourth of an inch of rise to the lineal yard of road, the force must be increased, or the load diminished, in a ratio or proportion varying at the rate of about one half, one third, one fourth, one eighth, and one ninth, etc.

" Such are the happy effects of a wise and extended policy, that, notwithstanding the expensive war in which this country has been engaged, more has actually been done in Great Britain, within the last twenty or thirty years, for the improvement of the highways, and in laying open the country by new and better lines of road, than was effected for centuries before that period. With such public improvements we presume to class the measure of the proposed railway from the city of Edinburgh and its port of Leith, calculated as it is to ramify through the various tracts of East Lothian, Berwickshire, Roxburgh and Selkirk shires, and to become, in time, a system of the greatest importance in its consequences to the advancement of the commerce and agriculture of this part of the kingdom. Under impressions of this kind, the noblemen and gentlemen who now come forward as promoters of this measure are actuated ; and with this in view, the reporter lays before them the accompanying survey, and will now endeavour to state the consideration which he has given the subject, by describing the several lines he has surveyed, and estimating the probable expense and advantages of the measure."

Mr. Stevenson then describes the proposed line, which he estimated at £52,000, and terminates his report by giving some remarks on the construction of railways, which are interesting as noticing the use of cast and malleable iron rails, and George Stephenson's experiments on locomotives.

" Construction of the Railway.

" In giving some general description or outline of the construction of the proposed railway, it may be observed, that the formation of railways, or roads of cast iron, is comparatively but a recent discovery, which, however, is likely to be attended with immense advantage to this commercial and agricultural country. From the great traffic to be expected upon the Edinburgh Railway, two sets of wheel-tracks will require to be laid,—one for the wagons or carriages coming to town, and another for those going to the country. This double railway, with the necessary allowance for driving-paths, etc., will occupy at least twenty feet of space in its cross-section, viz., four feet three inches for each set of tracks ; a space of four feet between the respective wagon-ways; and three feet nine inches on each side for a driving-path, fences, and gutters. The horse-paths, or spaces between the waggon-tracks of the railway, as proposed above, will be four feet three inches in breadth, or the width of the *square part* of the common cart axle, it being also a great advantage for the convenience of loading, etc., and for the stability of the railway, to have broad and rather low wagons. But from the general use to which this public railway is applicable, it may be found advisable to acquire even a greater breadth than twenty feet. The space between the tracks will be made up with stones, broken very small, and blinded or covered with gravel, as in the best description of road-making. The footpath for the drivers may be made with gravel, coal dust, pan ashes, or brick-dust, as may be found most convenient in the district of the railway.

" *Cast Iron Rails.*

"The cast iron tracks of the earlier railways were made flat, or about four inches in breadth, with a projecting ridge or *flange* upon the outer verge, and are technically called *plate rails*. But the reporter is led from his own observation, and the opinion of the following professional gentlemen obligingly communicated to him, viz., Mr. Wilson of Troon, Mr. Bald of Alloa, Mr. Landale of Charlestown, Mr. Grieve of Sheriff Hall, and Mr. Buddle of Newcastle, who are not only scientifically but practically conversant in this matter, to conclude that the plate rail not only induces greater friction, but is more exposed to have the wheels clogged and interrupted with gravel or small stones than that called the *edge rail*, which, in its best construction, of cast iron, consists of a bar of about 1½ inch in thickness or breadth, for the *seat* of the wheel, and of a depth corresponding to the weight to be carried. This bar is set upon edge instead of being laid flat. In this manner the edge rail presents less friction, and, weight for weight, is much stronger for the load than the plate rail; upon the same principle as, in modern carpentry, the beam is now set on edge, instead of being laid on its side as formerly. The Reporter is therefore to recommend an edge rail warranted to work with two tons, including the wagon, of the weight of 140 lb. per lineal yard of finished double railway. Lighter dimensions might indeed be found to answer; but for a public railway, the rails should be made of a greater strength than is barely sufficient for a given weight, as this cannot always be kept within bounds, or regulated to a nicety. The expense of a little additional weight of cast iron, in the first instance, will be greatly compensated in the end, by avoiding frequent repairs, and will thereby be amply repaid, while the expense of laying the road, and other contingencies, are much the same in the light as in the heavy rail. The mode of fixing is another point of great importance in the construction of a substantial railway. In the early practice of laying railways, the value of this new discovery was for a time lost to the public, owing to the intricacy and difficulty of this part of the design. Much trouble and expense have in this way been occasioned, in consequence of using, for the underground fixtures, soft and friable stones, liable to be acted upon by the alternate changes of the weather, from their being necessarily placed so near the surface. A method has been adopted of making the cross fixtures under ground, with bars wholly of cast iron, to which the rails are attached, with iron pins. Much, however, depends upon the nature and tenacity of the ground to be passed over. At the works of Lord Elgin and the Carron Company, the use of the sleeper or cross iron bar is laid aside, and other alterations are daily suggested as improvements, in the method of laying and fixing the rails, and also in the construction of the wheels and wagons. With regard to the construction of *cast iron rails*, they are, in general, made in the lengths of from three to four feet; but the reporter is inclined to think that the perfection of the cast-iron railway will be found to consist rather in shortening the rails very considerably than adopting even the shortest of those lengths; but this and similar matters will fall more properly to be matured in the practical details of the business.

" Malleable Iron Rails.

" One point, however, deserves particular notice here, as likely to be attended with the most important advantages to the railway system, which is the application of malleable iron instead of cast iron rails. Three miles and a half of this description of railway have been in use for about eight years on Lord Carlisle's works at Tindal Fell in Cumberland, where there are also two miles of cast iron rail; but the malleable iron road is found to answer the purpose in every respect better. Experiments with malleable iron rails have also been made at Mr. Taylor's works at Ayr and Sir John Hope's at Pinkie; and, upon the whole, this method, in the case of the Tindal Fell Railway, is not only considerably cheaper in the first cost than the cast iron railway, but is also much less liable to accident. In the use of malleable iron bars the joints of the railway are conveniently obtained, about twelve feet apart, and three pedestals are generally placed between each pair of joints.

" Locomotive Engine.

"Some of the most striking improvements in the system of railways are the patent inventions of Mr. Stephenson of Newcastle, particularly his *locomotive engine,* by which fifty tons of coal and upwards are at one load conveyed several miles along a railway by the power of steam."

Acting on the same general principles, Mr. Stevenson surveyed and reported on such lines as the " Montrose and Brechin Railway," the "Strathmore Railway," and the "East-Lothian Railway," which, as has been shown, embraced a large portion of the principal business part of Scotland. But at that time Scotland was not ready either to take up his enlarged views, or to find money to carry them out, and the prospectuses issued by the different Committees who zealously promoted these railway schemes did not meet sufficient support to enable the promoters to form Companies to apply to Parliament for their construction. We all know that in England, at a later date, our British Railway system was first inaugurated, but it is a fact that redounds greatly to Mr. Stevenson's credit as an engineer, that all of these Scottish lines, originally surveyed by him, have, with or without deviation, been now carried out.

Mr. Stevenson, in his researches for adapting railways

to the general communication of the country, had made a great advance in bringing the subject before the public; and he was requested to visit the coal districts in the north of England to advise as to establishing a railway between Stockton and Darlington, with extensions to the coal fields of Bishop-Auckland; which he did in 1819, meeting with Mr. Pease, Mr. Backhouse, and other influential men there, to whom, after making a survey, he reported on the Stockton and Darlington Railway.

In making these various researches, Mr. Stevenson was enabled to suggest many proposals which can only be regarded as valuable for the period at which they were made, but he gave many opinions, which undoubtedly have come wonderfully true in the history of railway communication.

The Right Honourable Sir John Sinclair, Bart., proposed, in 1823, certain queries to Mr. Stevenson relative to a proposal for the construction of an iron railway between the cities of London and Edinburgh, and the following is an extract from his reply, showing, that while he fully appreciated the value of *ship-canals*, he entertained the conviction that "iron railways" would become, as I have already said, the highway of the future.

"Regarding the practicability of such a scheme, it may be noticed that the late eminent James Watt entertained an idea of the eligibility and great advantage which might accrue to the public from the formation of a central and considerably elevated line of inland navigation constructed so as to ramify through the interior districts of England, and communicate with the principal manufacturing and populous towns in the kingdom.

"In any comprehensive view of a measure of this kind there

can be no doubt that an iron railway would not only be much more practicable, but more commodious and useful for general intercourse than a canal. And the comparative expenses of the two operations would probably be in the ratio of about one to eight in favour of the railway. Again, if the advantages of carriage by the railway and the *common road* be compared, it will be found that the proportion is at the rate of about one to seven, also in favour of the railway.

" The economy of carriage on the railway, when fully contrasted with that of the canal, is also much greater. It may now, indeed, be considered as a generally received opinion, that, unless for enabling sea-borne ships to pass from one side of the coast to another, so as to avoid a tedious or dangerous circumnavigation, the railway in every other case is preferable. It is at the same time to be noticed that when Mr. Watt suggested the idea of a central line of canal many years since, the railway system was then neither so well known nor so much acted upon as now."

Mr. Stevenson's belief that railways would ultimately be the general highways of the world, led him to regard with distrust their *immediate* introduction into Britain in absence of some public Act for their proper regulation, and accordingly, on 29th January 1825, he writes to Lord Melville in the following terms :—" It seems necessary at this time, even before any Act is proposed for a public railway, that a Committee of the House should take the subject of regulating the width according to the number of tracks, and perhaps the strength of rails and weight to be carried on four wheels, in a public Act, otherwise much confusion will ensue. It will be a great loss if these railways, like the common road, should require to be altered that they may communicate with each other.

"All the engineers I have spoken with, including Mr. Telford, agree in this. I have noticed it to Mr. Home Drummond and Mr. Gladstone.

"I put the specification of the bridge at Melville Castle in train before I left home."

Had it been possible to carry out the spirit of this suggestion, made at that early period, in an Act of the Legislature, I think, in the retrospect of much that took place during our "railway manias" and "railway company competitions," it might possibly have proved advantageous to the community.

The Highland and Agricultural Society of Scotland, which has ever been foremost to encourage everything that tends to the improvement of the country, regarded the introduction of railways as a matter of great importance, and considering it a subject that came legitimately within their province, offered, in 1818, a premium of fifty guineas for the best essay on the construction of railroads. Many competing treatises were given in, and the Society placed the whole of them in the hands of my father for his opinion and report on their merits, "together with such remarks of his own as he might judge useful." The result of his examination is given at great length in the Transactions of the Society,[1] accompanied by "notes," in which he makes several valuable suggestions. Before the period alluded to, the rails in use had been almost invari-

---

[1] The essays most favourably noticed are those of Mr. Alexander Scott, Mr. George Robertson, Mr. George Douglas, Mr. John Ruthven, Mr. James Dickson, Mr. James Walker (Carron), Mr. James Walker (Lauriston), Mr. John Fraser, Mr. John Wotherspoon, Mr. John Moore, and Mr. John Baird.

ably made of cast iron or timber; but my father in his notes says—"I have no hesitation in giving a decided preference to malleable iron formed into bars from twelve to twenty feet in length, with flat sides and parallel edges, or *in the simple state in which they come from the rolling-mills of the manufacturer."* He also recommends that they should be fixed into guides or chairs of iron supported on props placed at distances in no case exceeding three feet, and that they should be connected with a clamp-joint so as to preserve the whole strength of the material. It is not a little singular that this description, given about forty years ago, may, to use engineering phraseology, be not inaptly called a " specification of the permanent way" of our best railways at the present day.

I close this chapter by giving a letter which shows the value that George Stephenson attached to my father's researches on railways, while it is at the same time interesting as showing the very moderate estimate which the great Railway Engineer at that time entertained of the performance of the locomotive engine—a machine which was destined ultimately to become, under his skilful management, so important an agent in changing the inland communication of the whole civilised world :—

" KILLINGWORTH COLLIERY,
June 28, 1821.

" ROBERT STEVENSON, ESQ.

" SIR,—With this you will receive three copies of a specification of a patent malleable iron rail invented by John Birkinshaw of Bedlington, near Morpeth. The hints were got from your Report on Railways, which you were so kind as to send me by favour of Mr. Cookson some time ago. Your reference to Tindal Fell Railway led the inventor to make some experiments on malleable iron bars, the result of which convinced him of the superiority of the malleable over the cast iron—so much so, that he took out a patent. Those rails are so much liked in this neighbourhood, that I think in a short time they will do away the cast iron railways. They make a fine line for our engines, as there are so few joints compared with the other. I have lately started a new locomotive engine, with some improvements on the others which you saw. It has far surpassed my expectations. I am confident a railway on which my engines can work is far superior to a *canal.* On a long and favourable railway I would stent my engines to travel 60 miles per day with from 40 to 60 tons of goods. They would work nearly fourfold cheaper than horses where coals are not very costly. I merely make these observations, as I know you have been at more trouble than any man I know of in searching into the utility of railways, and I return you my sincere thanks for your favour by Mr. Cookson.

"If you should be in this neighbourhood, I hope you would not pass Killingworth Colliery, as I should be extremely glad if you could spend a day or two with me.—I am, Sir, yours most respectfully,

"G. STEPHENSON."

CHAPTER VIII.

HARBOURS AND RIVERS.

1811—1843.

THERE is scarcely a harbour or river in Scotland about which, at some time, Mr. Stevenson was not asked to give his advice. His opinion was also sought in England and Ireland, and he executed works of greater or less extent in many of the cases in which he was consulted.

We may select from his reports the names of Dundee, Aberdeen, Peterhead, Stonehaven, Granton, Fraserburgh, Ardrossan, Port-Patrick; the rivers Forth, Tay, Severn, Mersey, Dee, Ribble, Wear, Tees, and Erne, as among some of the many places in the United Kingdom where he was employed.

In a subsequent chapter extracts will be found illustrating Mr. Stevenson's views on various professional subjects, and from these it will be seen that he brought his large experience and study of the waves to bear advantageously and practically on his harbour engineering. He was, as will be gathered from the extracts, at an early period fully alive to the value of spending basins for tranquillising a harbour, and of the proper disposition of the covering piers, in reference to the line of exposure, so as to avoid throwing sea into the harbour's mouth, or causing it to heap up on coming in contact with the piers; while, as regards rivers, he was no less alive to the value of *backwater* in keeping open estuaries, and to the necessity of removing all obstructions to the free flow of the tide in river-navigation.

At an early date, for example, Mr. Stevenson and Mr. Price were jointly consulted as to the navigation of the Tees, and I am indebted to Mr. John Fowler of Stockton, the engineer to the Tees Navigation, for the following statement as to the result of that joint reference :—

" The Navigation Company consulted Mr. Stevenson and Mr. H. Price, who differed in opinion as to the general treatment of the river. Mr. Price recommended that it should be contracted by jetties, and Mr. Stevenson that the banks should be faced with continuous walls, stating as his reason for this recommendation, that ' to project numerous jetties into the river, I regard as inexpedient, being a dangerous encumbrance to navigation, and tending to disturb the currents and destroy the uniformity of the bottom.' The plan adopted by the Navigation Company was, however, that of Mr. Price ; and jetties were constructed on the river to a large extent," and Mr. Fowler adds, that " after a trial of twenty-seven years it was found that they were liable to all the objections that had been urged against them by Mr. Stevenson."

Accordingly, under Mr. Fowler's direction, the whole of the jetties have been removed.

One of the early harbour schemes in which my father

was engaged in England, was a harbour at Wallasey Pool, on the Mersey, in which he acted in conjunction with Telford and Nimmo. The following reports will show the nature and extent of work then contemplated as a commencement of the Birkenhead Docks, now so valuable an adjunct to the port of Liverpool. But at the early period of 1828, when the reports were written, the public were not prepared to entertain a scheme of improvement based on so great a scale. It included, as will be seen, not only the formation of a floating harbour at Wallasey on the Mersey, but the construction of a harbour at Helbre on the Dee, with a connecting ship canal between the two estuaries.

" To the Subscribers for the proposed Wet Docks at Wallasey Pool.

"Preliminary Report of Robert Stevenson and Alexander Nimmo, Civil Engineers, on the proposed improvements at Wallasey Pool.

"*Liverpool, Feby.* 23, 1828.—Having been requested to examine the situation of the Wallasey Pool with a view to discover how far additional accommodation might be obtained there for the increasing trade of the port of Liverpool, we did accordingly meet at Woodside on the 10th February 1828, and after examining the pool at high and low water, and the action of the tides on the northern edge of the Leasowe level, which we found to be overflowed at high water of the 16th and 17th and 18th February, with off-shore winds and moderate weather, we next examined the shore down to low water in that place called Mockbeggar Wharf, which we found to consist of turf and soft marl over a bottom of fine clay. We afterwards visited the western part of the level, which extends to the immediate vicinity of the estuary of the Dee, part of which we examined, also Helbre, Hoylake, and the Rock Channels, and directed certain surveys and levels to be taken for our further information, and though we have not yet obtained all the data requisite for forming estimates of the expense of improvement, we are generally of opinion as follows :—

"That this situation of Wallasey Pool affords, beyond doubt, the most favourable position in the vicinity of Liverpool for an extension of the accommodation of the shipping trade of the port, at a very moderate expense.

"The ground being level, the soil water-tight and of easy excavation, docks may be formed there of any extent. The bay in front between Seacombe and Woodside, though mostly shallow at present, affords the first place of shelter within the Mersey, and small vessels lie there out of the stream in perfect safety. It possesses a creek or channel which could easily be enlarged and deepened so as to form an outer tide harbour similar to the original harbour of Liverpool, but upon a greater scale, and for the scouring of which it would be easy to open up the tide in the pool to the extent of 250 acres, as far as Viners Embankment, and above that to any extent that may be thought desirable. This space having a deep creek through its whole extent forms a complete half-tide basin for facilitating the entrance into the Docks on either side, while on the shallow parts may be formed extensive timber-ponds. Works of masonry in this situation being out of the sea-way and of the stream of the tide, may be constructed with great economy; good building stones are to be found at Bidston Hill, and the whole soil is a brick earth.

"The situation possesses other advantages of access not so obvious, but which may eventually be of the greatest importance. The Leasowe level at the head of this pool extends as far as the river Dee, and touches the sea-shore at Mock Beacon, where indeed it is occasionally overflowed by the tide. In this direction it would be quite practicable to open a direct passage for ships into the Horse Channel, by excavating in marl and clay, only quite clear of the shifting sands which are found in all other parts of the Mersey and Dee. And towards the Dee a ship canal may easily be cut with its entrance either at Dawpool in Hoylake, or in a tide harbour which could be formed at Helbre, a position which affords many maritime advantages.

"That position has several good anchorages in its vicinity, three different passages to sea, and is only five miles from the floating light, the distance of which from Liverpool by Wallasey and Helbre is exactly the same as by the Rock Channel; and nine miles of it would be inland navigation, instead of an intricate passage among sandbanks, the whole of which inland navigation is an addition to the floating harbour.

"Having thus briefly shown the facilities possessed to seaward, we may next turn our attention to those connected with the inland navigation. It is evident that to the 'flats' which navigate the Duke's Canal, Mersey and Irwell, Ellesmere, Sankey, and Weaver Navigations, Wallasey Pool is just as accessible as the Docks of Liverpool, while by a canal to Helbre you communicate with the large navigation of the Dee, and the valuable mineral county of Flintshire; and if ever, as is extremely probable, the canal navigation should be brought nearer to Liverpool, the natural termination would be Tranmere or Wallasey Pool, between which a cut can be easily formed. By this means boats from the small canals in Staffordshire and the other inland counties can be brought down to the seaport and return their cargo without the trouble of transhipment,—an object, as being important to the proprietors of these canals, that there can be little doubt of their endeavouring to carry it into effect whenever the shipping can be accommodated on the Cheshire side.

"Although in the present state of our survey, and until we meet our eminent friend and colleague Mr. Telford, we are not prepared to enter into any detail of plans or estimates of the expense of these improvements, yet we are satisfied he will agree with us in opinion that the cost of even the most expensive will be greatly inferior to that of obtaining any important additional accommodation upon the Liverpool shore, which being almost entirely occupied already, we consider it impossible to obtain there at any expense sufficient room for the increasing trade; and we would conclude this preliminary report by recommending to the thriving and enlightened community of Liverpool to weigh well the advantages above alluded to, and the benefit of now extending their operations to the Cheshire shore.

"ROBERT STEVENSON.
ALEXANDER NIMMO."

" INTENDED SHIP CANAL between the RIVERS DEE and MERSEY.

" The REPORT of THOMAS TELFORD, ROBERT STEVENSON, and ALEX-ANDER NIMMO, Civil Engineers, recommending Two extensive new Sea Ports, etc., on the Rivers Dee and Mersey, adjacent to Liverpool, with a Floating Harbour or Ship Canal to connect them.

"The undersigned, having so far completed their land and water surveys as to enable them to speak with confidence upon the practicability of extending the accommodation for shipping to suit the rising demands of this great commercial emporium, beg leave to commence their report upon this important subject by describing the general outline of the proposed improvements, and then to proceed to discuss them in detail; but previous to this it is necessary to make a few preliminary remarks.

" *On the Estuaries of the Dee and Mersey.*

" In one or other of these must always continue to be the great port of the north-west of England, the preservation and improvement of which has become the more important since this last

century has added so much to the progress of manufacturing and
commercial enterprise, and to that extension of inland navigation, which has rendered Liverpool not only the great mart of the
north-west of Britain and of all Ireland, but nearly of the whole
western world.

"The chief feature of these estuaries is the extensive range
of sandbanks in their front, through which an intricate ship-
navigation has to be carried. These channels have been always
subject to variations, and are now only safely navigated by a careful system of pilotage.

"In the progress of our investigations, and feeling the great
importance of the measures we are about to recommend, we have
carefully inquired into the various changes which have taken
place on these banks, as far as can be collected from history or
inferred from observation, in order to be enabled to judge what is
likely to take place as to their future permanent condition.

"In the time of the Romans the Ribble seems to have been the
chief port of this district, and Ribchester is said to have been a
city as great as any out of Rome; the port was Poulton below
Preston, at the Neb of the Naze, so vastly inferior at the present
time to various situations on the Mersey and the Dee that it is
impossible not to admit that some extraordinary change has taken
place in their physical condition since that period. Tradition says
that the port of the Ribble was destroyed by an earthquake, and
also that there were tremendous inundations in Cheshire and
Lancashire about the termination of the Roman sway in Britain;
and various phenomena we have seen seem to point to some such
catastrophe.

"It is well known that in the Saxon times the river Dee was
an important navigation, and that Chester was then and for many
ages after the great port of the west, and for the connection with
Ireland, whilst the Mersey was little known, and Liverpool only
a fishing village.

"But in after times the port of Chester was so much obstructed by sandbanks in the upper portions that the city became inaccessible to vessels of large draught, and though serious efforts were made to remedy this evil, and have even partly accomplished it, yet the trade of the country was gradually transferred to Liverpool on the Mersey, which had become a place of considerable importance at the time of the Revolution, and had been created an independent port: before, it was only a creek of Chester.

"In our inquiries into the early state of the navigations of the Dee and Mersey, the oldest chart we have found of any authority is that of Grenville Collins, in 1690. It is dedicated to King William, to whom he acted as pilot on his expedition to Ireland; and as that army embarked from Hoylake, as also that of the year before under General Schomberg, and as Collins was officially employed in making charts of the coast, there can be no doubt that, though rude, it conveys, as far as it goes, an authentic representation of the state of navigation at that time.

"The roadstead of Hoylake was then spacious and deep, with five fathoms into it, and seven fathoms inside, from one half to three quarters of a mile wide, and covered by the Hoyle Sand, which was then one solid bank without any swash or opening across it, and was dry at neap tides as far as opposite the Point of Air and beyond.

"The Dove Point then projected a mile and three-quarters from the shore, separating Hoylake from the Rock Channel, which was then nearly dry at low water as far as Mockbeggar, between which and Burbo Sand there was only one quarter fathom, and between Dove Point and Burbo only two fathoms.

"The large vessels which at that time belonged to Liverpool put out part of their lading in Hoylake until they were light enough to sail over the flats to Liverpool.

"The union of Hoylake and the Rock Channel formed, as at present, the principal passage to sea, called the Horse Channel, then a fair opening with three to seven fathoms, but considerably to the eastward of the present channel of that name; for Collins's sailing mark through it was Mockbeggar Hall upon the Banquetting-House in Bidston, would mark the present Spencer's Gut as having been the channel.   The north spit did not then exist, or rather was part of the Hoyle bank; and the Beggar's Patch seems to have been the extremity of Dove Point.   The Formby Channel was said to have three fathoms on the bar, but was not buoyed or beaconed, therefore not used.

"The Chester bar had nine feet least water; and Wild Road is marked as good anchorage, much used in the coal trade.   About 1760, published in 1776, we have the Survey of Mackenzie, who was employed by the Admiralty to make charts of the western coasts of Britain, which are still in high reputation.

"At this time Hoylake continued to be a good roadstead, though greatly altered; the depth at entrance was only two fathoms, eight fathoms in the middle, the width only three furlongs, and its length had diminished at least a mile.   A passage was opened from the Rock Channel across to Dove Point into Hoylake, and across the east end of Hoyle Sand, with four to eight fathoms, forming the present Horse Channel.

"On this chart we also perceive the beginning of another opening across the Hoyle Sand, now called Helbre Swash, then dry at low water at each end, having three fathoms in the middle, now a deep and fair channel with seven to nine fathoms, and two and a half least water at its mouth.

"Since the opening of this channel or swash little or no tide sets through the Hoylake, which is gradually closing up, and now used only for small craft.

"The existence of Hoylake was of material importance to Liverpool and also to the Dee, for vessels could run there at any time; the entrance to it was marked by leading lights in the middle of last century, one of the first applications of reflecting lights to the purposes of navigation; they are now of little use, as the sand has shifted to the eastward, and the entrance is nearly dry at low water.

"The Rock Channel seems to have undergone a very important change by the time of Mackenzie's survey. We have observed that in Collins's time, 1690, it was dry at low water as far nearly as Mockbeggar. Although this is still nearly the case at the Perch at low tides, it is opened below that in a material degree. In the space of seventy years the channel had deepened to have three or four fathoms in Wallasey Hole; also between Mockbeggar Wharf and the north bank, which was dry at low water; and a channel had opened across Dove Point, with two and three fathoms, into Hoylake, and from thence across the east end of Hoyle, forming the present Horse Channel, as before described, with four to eight fathoms out to sea. On the other hand, the sand from this deepening had been carried down to seaward, forming a complete shoal across the original Horse Channel of Collins's time, in whose sailing-line is marked a depth of four feet only, and this shoal connected with that called the Beggar's Patch, and thence with the spit or flat along the west side of the Horse Channel, on which was six feet water. This last channel was direct and fair, with five to eight fathoms, and previous to the publication of Mackenzie's chart, but after the time of his survey, was marked by two lighthouses at Leasowe shore, and subsequently by that on Bidston Hill under the direction of Captain Hutchinson, as was also the entrance into Hoylake by the two lights near Meols, as before described.

"The Formby Channel is marked as deep upon Mackenzie's chart, with four fathoms at the entrance, and between Taylor's Bank and Middle Patch two fathoms; there is now only five feet over the flats at low water at its entrance, and it was buoyed in at Mackenzie's time; but, though the deepest channel to Liverpool, it is, from its intricacy and instability, still very little used for navigation.

"Lieutenant Evans published a survey of the Liverpool and Chester rivers, with a book of sailing directions, which is in good repute. We have preferred the chart by Mr. Thomas in 1813, made by order of the Lords of the Admiralty, for the purpose of comparison with the several before mentioned surveys, as more minute in detail.

"At the time of this survey, fifty years after that of Mackenzie, Hoylake had diminished in breadth to one furlong; the depth at the entrance was three to seven feet; four fathoms near the Red Stones; since that time it is still shallowing, and now may be walked across at low water, from Dove Point to East Hoyle; so that this roadstead may be considered as lost.

"Helbre Swash had opened to half a mile wide, with six or eight fathoms water, but with a shoal at its entrance of one fathom; there are now two fathoms and a half through that entrance.

"The Brazil or North Bank had extended dry, at low water, as far as Spencer's Gut Buoy, and the North Spit or four feet flats had extended into the Horse Channel across the line of sea lights, thereby forcing that channel further into Hoyle Bank. The lower part of the Rock Channel had enlarged by the formation of a passage on each side of the Beggar's Patch.

"The entrance to Formby Channel had very much altered since Mackenzie's time, and, though better marked, still continued to be little frequented. The floating light placed opposite Helbre Swash and the Horse Channel, outside of all the banks, has made a great improvement in the access from the seaward in that direction.

"The Rock Channel, from these circumstances, continues to be the main passage to and from the harbour of Liverpool, but it is only provided with day marks, and though well buoyed cannot be navigated by night; being very narrow, and having banks in its middle, it is difficult for vessels to beat through with foul winds in one tide, and as there is no secure anchorage, frequent delays and losses take place in this part of the navigation.

"Within the harbour of Liverpool or in the river Mersey the principal places of anchorage are—

"1st, Abreast the town.

"2d, Off the Magazines, which is used by the outward-bound vessels.

"3d, Up the river in Sloyne Roads, or Broombro Pool, which is almost confined to vessels under quarantine.

"In the two first-mentioned anchorages a great sea tumbles in, with NE. gales, and this, with the rapid tide and bad holding ground, causes vessels to drift, even with two anchors down, so that it is necessary for all the merchant vessels, as soon as the tide serves, to proceed into dock and remain there until a favourable opportunity occurs of putting to sea, so as to get through the Rock and Horse Channels with daylight; hence a considerable accumulation of vessels within the docks at all times, but especially when there has been a continuance of northerly and westerly winds, and which has made it necessary to look now for additional accommodation on the opposite shore of Wallasey Pool.

" Proposed establishment at Wallasey.

"Small craft find good shelter on the banks at the mouth of Wallasey Pool, being there out of the stream, and land-locked by the Point of Seacombe.

"The steamers also, to which dispatch is of moment, moor along this shore, and if there was more room in Wallasey Pool it would decidedly be the best anchorage about Liverpool.

"Wallasey Creek runs nearly for two miles from the Mersey, where it is stopped by an embankment, through which the waters of 3000 acres of marsh land pass by a tunnel. The pool below the embankment covers nearly 250 acres at spring-tides, and by its backwater maintains a channel through the creek down to low water springs, and with seventeen feet at high water springs as far up as the embankment.

" Previous to the embankment it is certain that this creek was materially deeper. On Mackenzie's chart, opposite to its mouth, there are twenty fathoms marked, being much more than anywhere within the Mersey at present, and a bottom of rock. This channel would therefore be restored by any considerable addition to the backwater; and at all events, if the lower parts of the creek were opened by dredging, and, by a power of scouring it, low water obtained, a safer inlet for vessels to run to would be acquired than at present exists anywhere in the neighbourhood of Liverpool.

" On the south side of the creek, between Woodside Ferry and Bridge End, there is a bottom of sandstone rock, but this ceases at Bridge End Creek; and above that place the shore is composed of firm clay, fit for brick making, to a depth at least of thirty feet, in which excavation for docks and basins could be carried on with great facility.

"Upon the attention being directed to Wallasey Pool as a commercial station, it will appear at first view obvious that an entrance might be made along the low ground which extends from it to the sea shore at Leasowe, by which a direct passage to sea might be obtained, and the insecurities and dangers of the bar and banks of the Rock Channel be avoided; but the objections to such an entrance are, that the channel outside affords no safe anchorage, and the cut would be exposed directly to the stroke of the sea, and if protected by piers their construction would not only be expensive, but might also materially alter the channel along shore.

"But the ground continues equally favourable to the westward as far as Hoylake and the Dee below the hill of the Grange. The shore is skirted by a narrow belt of sandhills, through which however there would be no great difficulty in making a passage into the tideway. Here it is important to remark that the Helbre Swash opens a deep and fair channel, well sheltered by banks on each side, and only five miles in extent to the floating light, which is in a direct line with it.

"This channel has been formed within the last century, and readily accounts for the deterioration of Hoylake; it now carries down-most of the ebb of the Dee, and is likely to improve still more, having deepened materially since Thomas's survey in 1813.

"Through all the vicissitudes we have traced there has been deep water and good anchorage at the point of Helbre Island; and as that situation affords solid rock for every sort of construction, there can be no risk of the permanency of any work that may be established there.

"Sea-locks constructed at Helbre would be protected against the prevailing westerly gales by the island itself, against the northerly by the bank of East Hoyle; and they may be connected to the mainland by banks formed across the strand, which is mostly dry at high water of neap tides; and by means of these banks a pond of sixty-four acres may be enclosed, which, being filled at spring tides, may be employed for the purposes of scouring and keeping open the harbour and its entrance, and as a reservoir for a ship canal from thence to the shore, and along the low ground to Wallasey Pool. Such a canal, of large dimensions, and seven miles long, will be one continued floating harbour, which may be carried to a great extent in various directions and on the same level.

"Independent of Helbre Swash two other channels for ships passing to sea unite at that position; one, the original Hoylake, still sufficiently navigable at high water; the other, the passage by Wild Road and Chester Bar, greatly superior in safety and permanency to that of the Formby Channel; for in all the successive charts little or no change seems to have taken place on that bar, which continues to have nine feet at low water, with a rise of thirty. The great extent of ebb-tide from the Dee (being quite as extensive an estuary as that of the Mersey) must always keep one or other of those channels or all of them open, so that ships may sail from Helbre in almost every wind; and if necessary to beat out, a vessel starting from Helbre with the first of the ebb down the Swash will be at the floating light and clear of the banks before another from Liverpool can get round the Rock Perch.

"To persons at all acquainted with the navigation to Liverpool it must be quite unnecessary to point out the benefit of this proposed arrangement, which, while it preserves all the advantages of communicating with the Mersey, and the extensive inland navigations connected therewith, affords a new passage to and from the sea, by means of the Dee, by which both the distance and dangers of an intricate navigation will be wholly avoided.

"An important advantage obtained by this plan is, that the proposed entrance at Helbre is within the jurisdiction of the port of Chester, of which it is recorded as a creek in Sir Matthew Hale's Treatise *De portibus maris;* and business done there or upon its waters, even as far as Wallasey Pool, being within the port of Chester, will have to pay the dues at that port; and unless ships and goods lock into the Mersey they are exempted from the dues of Liverpool. The facility of construction is so great that a moderate charge for dues will be a sufficient remuneration for the capital required. The ground on either side of the canal is singularly suitable to be appropriated to any kind of establishment connected with shipping, and there can be no doubt that it will be so employed even by private speculation; but in so extensive a scheme as we propose it will be advisable for the promoters of the measure at once to establish a set of docks and warehouses of the most perfect description, as has been done in all the docks which have been constructed in and adjacent to London, and we have accordingly designed a set of such warehouses and yards as part of the plan.

"Details of the Plan.

"Commencing at the river Mersey, we propose to dredge out and widen Wallasey Creek at least to the depth of three feet under low water of spring tides, being four feet below the sill of Prince's

Dock, and this for 200 feet in width up as far as the entrance into the basins; to lay the sill of the greater entrance lock at that level, also the sill of the basin of the barge lock. The barge entrance lock to have a lift of ten feet; the ship lock four feet; so as to give the same water when the gates are opened as into the Prince's Dock. The side of Wallasey Creek will be quayed for four hundred yards below the entrance of the dock, to facilitate transporting vessels into and out of the basins.

"The tide basin is 1000 yards long, and 100 yards wide in the middle, curving on the north side towards the locks at each end, the south side receding 100 feet, so as to give berthage to timber vessels, and in the front of them a sloping wharf and bonding yards for timber; a line of barge canal between these yards and the warehouses on the main dock will facilitate the removal of the timber without interfering with the shipping.

"The entrance lock into this basin from the tideway will be fifty feet wide, the entrance wing walls widening gradually to 100 feet, to afford easy access to the shipping when both gates are thrown open. At low water, neaps, or half tide, two or three vessels may pass at a time. The upper lock between this basin and the canal to be double; one large lock, forty-five feet wide and 160 feet long, for great ships, and another, twenty-five feet wide, for smaller vessels, with gates at each end, pointing both to land and seawards. These locks to rise to four feet below the old dock sill of Liverpool, and thus to have twenty-two feet water in the canal on the level of an eighteen feet tide, which we propose to make the surface level of the canal.

"The ship dock parallel to this basin will be 400 yards long and 100 wide, with warehouses on each side, supported by iron pillars, so as to form a covered wharf, as at the St. Katherine's Dock in London; behind these warehouses a parallel barge canal fit for river flats, forty feet wide, which will, as in Holland, be found

T

a singular convenience. These canals communicate with a dock and basin for flats only, whence the barges may be let down into the creek during the ebb; and as they navigate at the lowest water they will be ready to pass up the Mersey with the first of the flood; and in like manner, coming down with the last of the ebb, will get into the pool and enter the dock without losing a tide. Ships from the Mersey, in like manner, may enter the basin with half-flood, and be ready to proceed down the Swash with the first of the ebb.

"The flat marsh by the Boilers Yards is well adapted for this establishment, but as the ground beyond is high for some distance we propose the canal to be 124 feet only at water surface for 1000 yards from the locks, and to be lined with a stone wall on each side, so that this space will, in fact, be also a dock. Afterwards the marsh widens, and here is a favourable place for another entrance basin and dock, if necessary. From this point we propose to continue the canal with sloping banks, the bottom to be four feet under the level of the old dock sill, and 163 feet in width at the surface of the water, which will be twenty-two feet in depth.

"The canal proceeds at first in the direction of the Leasowe Lighthouse, and approaches within half a mile of the shore, and about the same distance north of the village of Moreton, and then turns to the westward, keeping half a mile inland from the villages of Great and Little Meols through Newton Car, where it turns off to Helbre Island, and enters the strand about half a mile above the hotel; across the strand it is carried by embankments to the upper end of Helbre Island. A large breadth is allowed for the embankment on the sea-side, with facing mound of stone from the rocky point near the Red Stone to within 600 feet of the Point of Helbre. The head of this pier to be of rough stone, rounded off, and carefully paved. A pier head is to be built in Helbre of 300 feet in length, leaving an opening of 300 feet into the tide harbour, which is fifty acres in extent, and to be cleared to at least low water of a spring-tide, and preserved of that depth by scouring.

"A quay wall is to be constructed of hewn stone along the Helbre Island from the pier-head 600 yards to the tide lock, which is to be fifty feet wide, as at Wallasey; another tide lock of similar dimensions on the north side of the harbour. The north pier is only intended to be of rough stone; but a short covering pier will be made to protect that lock and facilitate the entry of ships. Above these locks the canal is to be formed into a tide basin of 500 yards in length, the level of which may be kept at that of the tide of the day; and at the upper end are two parallel canal locks, as at Wallasey, with gates pointing to the sea and land at each end, as the tide will occasionally rise higher than the level of the water in the canal.

"From Helbre Island to the Middle Helbre, thence to the Eye, and from that to the shore at Kirby Church, an enbankment and road will be carried along the ridge and made water-tight. By this and the canal a pond, as has already been described, will be enclosed, of 640 acres, which will fill at spring-tides to the depth of nine feet, containing 3,000,000 of cubic yards, and may be all emptied for the purpose of scouring the outer harbour; but at the latter part of the spring tides it will be advisable to fill this pond as a reservoir for lockage water, for which purpose it may be drawn down three feet to the canal level, and will hold 1200 locks-full for ship lockage at each end, and, if necessary, 1000 more locks-full may be drawn off without any material inconvenience to the navigation.

"We now subjoin an estimate of what we conceive will be the expense of completing these works, including an extensive range of warehouses on each side of the dock at Wallasey Pool, and of enclosed timber yards along the tide basin; and for all the items we have made a liberal provision.

"*Estimate.*

Excavations in Wallasey Creek and Helbre Harbour, also in the
  Locks, Basins, and Canal to Helbre, and Barge Canal and Basins,   £436,017
Quay Walls on Creek, Basins, Locks, and Canal at Wallasey Pool, .   230,100
Bridges and Tunnels, ...   38,000
Piers and Quays Walls, Helbre Harbour, ...   95,100
Locks, Dams, and Culverts, Helbre Harbour, ...   111,000
Warehouses at Wallasey Pool, Inclosure Walls, and Paving, . .   183,000
Purchase of Land, ...   125,000
For Surveys, Act of Parliament, Law Expenses, Superintendents',
  Lock-keepers', and other Offices, etc., and Contingencies on Works,
  Fifteen per Cent., ...   182,731
                                                                  _____
                                                                  £1,400,948

" For the above sum a floating harbour will be obtained of seven
miles in length, capable of indefinite enlargement, with extensive
warehouse accommodation, and with a sea-port at either end on
the two separate estuaries.   That this is not too great for the wants
of the country will be at once admitted by those who consider the
vast extent of shipping usually moored in the Thames, notwithstanding all its docks; the total inapplicability of the rivers Mersey
or Dee to such a purpose; and the confined space which even
the docks of Liverpool can afford for the accommodation of a trade
now hardly inferior to that of the metropolis, and certainly and
rapidly increasing.

" THOS. TELFORD.

ROBT. STEVENSON.

ALEXANDER NIMMO.

"LONDON, 16*th May* 1828."

"FURTHER REPORT respecting the proposed two new Ports, etc., on
  the Rivers Dee and Mersey, adjacent to Liverpool.

" In the foregoing report we have shown the form and expense
of this establishment when completed upon an extensive and perfect plan.   At the commencement, however, of so great an under-

taking it is not to be expected that all the conveniences we have proposed can be immediately required; a considerable portion may therefore be deferred until the wants and increasing demands of trade shall show them to be necessary. In the meantime the essential parts of the improvement may be effected, with a smaller expenditure of capital, so as to obtain all that safety and facility of access which we have shown to be leading features of this plan.

" We have proposed to make the canal from Wallasey to Helbre wide enough for three great ships, so as to admit of part of it being used as a floating harbour, still leaving room for navigation; but for navigation alone it will be quite enough to adopt the dimensions of the Caledonian Canal, viz., 120 feet at surface, and if the trade should increase so as to require it, instead of widening it, a parallel canal may hereafter be made, with a bank and two towing-paths between, leaving the whole of the opposite banks applicable to berthage and commercial establishments. The same locks will serve at either end, and the transporting of ships be greatly facilitated; and the construction of this canal, or repair of the other, may be effected without any interruption to the navigation by such an arrangement. Again, the double locks at the Wallasey end of the canal, intended for the greater dispatch of business, may very well be deferred for the present, and the entrance basin made of smaller dimensions. The ship dock there may at first be made as a part of the canal, and quayed on one side only, and afterwards widened and completed when wanted. The half tide dock may be dispensed with by enlarging the barge tide dock so as to serve also for ships, and the quay walling of the pool and of the first mile of the canal may also be deferred. The warehouses at Wallasey dock may be dispensed with at first, or left to individual capital; but it will be highly proper to secure a sufficient quantity of land to enable all these improvements to be undertaken at some future period. We do not deem it advisable to give up the enlargement and deepening of the entrance of Wallasey Pool, as on that depends much of the utility of the plan in giving access to vessels at low tides ; and for a similar reason we would preserve all the works proposed for the harbour at Helbre Island. Upon this modified plan the expense, as below, will be £734,163.

" THOS. TELFORD.
ROBT. STEVENSON.
ALEX. NIMMO.

" CHESTER, *July* 14, 1828.

" *Estimate.*

Excavating Tide Basin, Barge Dock, and half of Ship Dock, at Wallasey End, ...	£25,000
Walling along the Pool, from Brassey's Works, also the Barge Dock and one side of Ship Dock, ...	31,500
Ship Lock, Barge Lock, and Tide Gates for Basin, and two Swivel Bridges, ...	36,500
Dredging Wallasey Creek, as before, ...	20,000
Land and Damages, ...	51,000
	£164,000
Fifteen per Cent. Contingencies, . .	24,485
For Wallasey End, ...	£188,485
Excavating Canal, ...	£207,403
Bridges and Tunnel, ...	22,000
Land and Damages, ...	27,000
	£256,403
Fifteen per Cent., ...	38,460
For the Canal, ...	£294,863
Pier and Quay Walls from Helbre, as before, ...	£95,100
Locks, Dams, and Culverts, do., ...	111,000
Excavation in Harbour, ...	10,000
Strand and Damages on Isle, ...	2,000
Fifteen per Cent., . . .	32,715
	£250,815
GENERAL TOTAL, . .	£734,163 "

I have given the Reports of the three Engineers to whom this question was remitted, to show the very comprehensive view they took of the important subject referred for their opinion; and it is almost unnecessary to tell professional readers that after a lapse of nearly a quarter of a century the embryo but comprehensive proposal of Telford, Stevenson, and Nimmo resulted in the modified but still large Birkenhead Dock scheme of J. M. Rendel.

The original design for the improvement of the Tay was made by Messrs. Robert and Alan Stevenson, in 1833, and in connection with my father's life a short account of the works may be desirable as illustrating his practice in River Engineering in the Tay and other rivers.

The river Tay, with its numerous tributaries, receives the drainage water of a district of Scotland amounting to 2283 square miles, as measured on Arrowsmith's map. Its *mean* discharge has been ascertained to be 274,000 cubic feet, or 7645 tons of water per minute. It is navigable as far as Perth, which is twenty-two miles from Dundee and thirty-two from the German Ocean.

Before the commencement of the works, certain ridges, called "fords," stretched across the bed of the river, at different points between Perth and Newburgh, and obstructed the passage to such a degree that vessels drawing from ten to eleven feet could not, during the highest tides, make their way up to Perth without great difficulty. The depth of water on these fords varied from one foot nine inches to two feet six inches at low, and eleven feet nine inches to fourteen feet at high water of spring tides ; so that the regulating navigable depth, under the most favourable circumstances, could not be reckoned at more than eleven feet. The chief disadvantage experienced by vessels in the unimproved state of the river was the risk of their being detained by grounding, or being otherwise obstructed at these defective places, so as to lose the tide at Perth,—a misfortune which, at times when the tides were falling from springs to neaps, often led to the necessity either of lightening the vessel, or of detaining her till the succeeding springs afforded sufficient depth for passing the fords. The great object aimed at, therefore, was to remove every cause of detention, and facilitate the propagation of the tidal wave in the upper part of the river, so that inward-bound vessels might take the first of the flood to enable them to reach Perth in one tide. Nor was it, indeed, less important to remove every obstacle that might prevent outward-bound vessels from reaching Newburgh, and the more open and deep parts of the navigation before low water of the tide with which they left Perth.

The works undertaken by the Harbour Commissioners of Perth for the purpose of remedying the evils alluded to, and which extended over six working seasons, may be briefly described as follows :—

1*st*, The fords, and many intermediate shallows, were deepened by steam dredging ; and the system of harrowing was employed in some of the softer banks in the lower part of the river. Many large detached boulders and "fishing cairns," which obstructed the passage of vessels, were also removed.

2*d*, Three subsidiary channels, or offshoots from the main stream, at Sleepless, Darry, and Balhepburn islands, were shut up by embankments formed of the produce of the dredging, so as to confine the whole of the water to the navigable channel, and the banks of the navigable channel were widened to receive the additional quantity of water which they had to discharge.

3*d*, In some places the banks on either side of the river beyond low water mark, where much contracted, were excavated, in order to equalise the currents, by allowing sufficient space for the free passage of the water; and this was more especially done on the shores opposite Sleepless and Darry islands, where the shutting up of the secondary channels rendered it more necessary.

The benefit to the navigation in consequence of the completion of these works was of a twofold kind; for not only was the depth of water materially increased by actual deepening of the water-way, and the removal of numerous obstructions from the bed of the river, but a clearer and a freer passage was made for the flow of the tide, which begins to rise at Perth much sooner than before; and as the time of high water is unaltered, the advantages of increased depth due to the presence of the tide is proportionally increased throughout the whole range of the navigation; or, in other words, the *duration of tidal influence has been prolonged.*

The depths at the shallowest places were pretty nearly equalised, being five feet at low and fifteen feet at high water, of ordinary spring tides, instead, as formerly, of one foot nine inches at low and eleven feet at high water. Steamers of small draught of water can now therefore ply at *low water*, and vessels drawing fourteen feet can now come up to Perth in *one tide* with ease and safety.

CHAPTER IX.

PRESERVATION OF TIMBER.

1808—1843.

IN 1808 Mr. Stevenson was the discoverer of the *Limnoria terebrans*, that small but sure destroyer of timber structures exposed to the action of the sea, and forwarded specimens of the insect and of the timber it had destroyed to Dr. Leach, the eminent naturalist, of the British Museum, who, in 1811, announced it as a "new and highly interesting species which had been sent to him by his friend Robert Stevenson, Civil Engineer," and assigned to it the name of *Limnoria terebrans* (*Linnean Trans.*, vol. xi. p. 37, and *Edinburgh Encylopædia*, vol. vii. p. 433).

The *Teredo navalis*, which is a larger and even more destructive enemy, is happily not so prevalent in northern seas as the *Limnoria*.

So impressed was Mr. Stevenson with the importance of his discovery as affecting marine engineering, and especially harbour works, that he resolved to establish a train of systematic experiments by exposing the timber of different trees to the action of salt water, and it occurred to him that no situation could be more suitable for such observations than the Bell Rock, where the specimens would not only be fully exposed to the sea, and free from any interference, but would be strictly watched and minutely reported on by the light-keepers. He further conceived it proper, in the interests of the navy, to take the Admiralty into his counsels, and he accordingly communicated his intention to that Board, with the result that many of the specimens of timber experimented on were sent from Woolwich dockyard, and the results of the trials were from time to time communicated to the Admiralty.

The different blocks of timber under trial were treenailed to the rock, and the experiments extended over a period of nearly thirty years. They clearly proved that teak, African oak, English and American oak, mahogany, beech, ash, elm, and the different varieties of pine, were found sooner or later to become a prey to the *Limnoria*. Greenheart oak was alone found to withstand their attacks, and even this timber was ultimately not entirely unaffected.

The result of these valuable experiments is given in the following Table :—

TABLE showing the different kinds of Timber which were exposed to the attacks of the *Limnoria terebrans* at the Bell Rock in 1814, 1821, 1837, 1843, with their durabilities.

Kind of Timber.	Decay first observed.		Unsound and quite decayed.		Quite sound for		Remarks.
	yrs.	mo.	yrs.	mo.	yrs.	mo.	
Greenheart,[1]	..		..		19	0	[1] Affected in one corner.
Teak-wood,	..		..		13	0	
Beef-wood,	..		..		13	0	
Treenail of Bullet-wood,	..		..		5	0	[2] A little holed at one end under-
Beech, Payne's patent pro.,[2]	10	7	..		..		neath.
Teak-wood,[3]	5	6	..		..		[3] Nearly sound 7½ years after being
African Oak,[4]	5	6					laid down.
Do. do.	4	11	10	0	..		[4] Nearly sound 7½ years after being
English Oak, kyanised,	4	7	10	0			laid down.
Teak-wood,	4	7	12	0			
American Oak, kyanised,[5]	4	3			..		[5] Decaying, but slowly, 5 years and
British Ash,	3	0	5	0			7 months after being laid down.
Scotch Elm,	3	0	5	0			
Ash,	2	11	4	3			
English Elm,	2	11	4	7			
Plane Tree,[6]	2	11	..		..		[6] Decaying, but slowly, 5 years and
American Oak,	2	11	4	7			7 months after being laid down.
Baltic Red Pine,[7]	2	9	4	3			[7] A good deal decayed when first
English Oak,	2	4	4	7	..		observed.
Scotch Oak,[8]	2	4			..		[8] Much decayed when first ob-
Baltic Oak,	2	4	4	3			served.
Norway Fir,	2	4	3	1	..		
Baltic Red Pine, kyanised,	2	4	4	7			
Pitch Pine,	2	4	4	3			
American Yellow Pine,	2	4	3	7			
American Red Pine,	2	4	3	1	..		
Do. do., kyanised,	2	4	4	7	..		
Larch,	2	4	4	3	..		
Honduras Mahogany,[9]	2	1	..		..		[9] Nearly sound 3½ years after being
Beech,	1	9	3	1			laid down. Washed away 6
American Elm,	1	9	3	1			months later.
Treenail of Locust,	..		5	0	3	0	
British Oak,	1	6	5	0	..		
American Oak,	1	6	5	0			
Plane Tree,	1	6	5	0			
Honduras Teak treenails,	1	6	5	0			
Beech,	1	6	5	0			
Scotch Fir, teak treenails,	1	6	3	0			
Do. from Lanarkshire,	1	6	3	0			
Do. do.	1	6	3	0			
Do. Locust treenails,	1	6	3	0			
Memel Fir,	1	6	5	0			
Pitch Pine,[10]	1	6	2	6	..		[10] Going fast when first observed.
English Oak,	1	1	3	1			
Italian Oak,	1	1	3	6			
Dantzic Oak,	1	1	2	6			
English Elm,	1	1	1	6			
Canada Rock Elm,	1	1	1	6			
Cedar of Lebanon,	1	1	2	6			
Riga Fir,	1	1	1	6			
Dantzic Fir,	1	1	1	6			
Virginia Pine,	1	1	1	6			
Yellow Pine,[11]	1	1	1	6	..		[11] A good deal gone 18 months after
Red Pine,	1	1	1	6			being laid down. Swept away
Cawdie Pine,[12]	1	1	1	6	..		by the sea 7 months afterwards.
Polish Larch,[13]	1	1	1	6	..		[12] A good deal decayed when first
Birch, Payne's patent pro.,	0	10	1	10	..		observed.
American Locust treenails,	0	8	3	0	..		[13] Going fast when first observed.

Mr. Stevenson seems to have formed an opinion that the best preservative against decay was charring the timber, as recommended in the following extract from a report, made in 1811, to the Trustees of Montrose Bridge :—

" The changeableness of climate to which the northern parts of this island are subject renders edifices of timber more liable to decay here than perhaps in any other country in Europe. But the bridge at Montrose is curiously circumstanced; for while it unavoidably exposes a great surface of timber to the action of the weather, some of the wooden piers are immersed twenty-two feet in the water, where they are attacked by a destructive marine worm. Some of the woodwork at the Bell Rock was infested with the same species of animal which preys upon the wooden pier at Montrose. In some of the temporary works there, as in the beams laid for carrying the railway over the inequalities of the rock, the timber was so much wormed that some logs measuring one foot when laid down would not square to more than nine inches at the end of three years. The beams which supported the wooden house for the accommodation of the artificers while the lighthouse was erecting escaped almost untouched, having been slightly charred, but the reporter, when inspecting the Bell Rock works this year, found that these worms are making some impression upon the ends of the supports resting on the rock where the charring could not take effect. The reporter is therefore of opinion that there is no better defence against the effects of this animal than slightly charring the timber, and he would recommend the practice at the bridge of Montrose wherever it can be applied. The operation of charring at the Bell Rock was performed by previously scraping off the adhering matter upon the logs and laying the skin of the wood open, and tar was applied to promote the combustion. Charcoal, besides being tasteless and inodorous, possesses some very curious properties in its action upon vegetable and animal substances, which may not only render it insipid, but even offensive to this insect. For those parts between the high-water mark and the roadway it will be enough to scrape the timber and lay it over with hot tar."

I need hardly say that this advice would perhaps not have been given at the present day, when even creosote has been found to delay, though not to act as a perfect defence against, the ravages of the *Limnoria*.[1]

PRESERVATION OF IRON.

At a more recent period Mr. Stevenson experimented at the Bell Rock Lighthouse in the same way on twenty-five different kinds of malleable iron, with the result that all of them were soon affected, and that galvanised specimens resisted oxidation from three to four years, after which the chemical action went on as quickly as in the others.

[1] Notice of the Ravages of *Limnoria terebrans* on Creosoted Timber.—*Proceedings of the Royal Society of Edinburgh*, vol. iv. and vol. viii.

CHAPTER X.

BRIDGES.

1811—1833.

Marykirk, Annan, Stirling, and Hutcheson stone bridges—High-level bridge for Newcastle—Timber bridge of built planks—Winch Chain Bridge—American bridges of suspension—Runcorn Bridge—Menai Chain Bridge—New form of suspension bridge.

MR. STEVENSON'S stone bridges over the North Esk at Marykirk, and the Nith at Annan (Plate VI.), are good specimens of road bridges of moderate extent; and his bridge over the Forth at Stirling, and Hutcheson Bridge over the Clyde at Glasgow (Plate VII.), are structures of a larger class.

Of the latter, Mr. Fenwick, of the Royal Military Academy, Woolwich, in the preface to his work on the *Mechanics of Construction*, published in 1861, says,— "The London and Waterloo Bridges, in the metropolis, which rank among the finest structures of the *elliptical arch*, and Stevenson's Hutcheson Bridge at Glasgow, which is one of the best specimens of the *segmental arch*, together with many others, have supplied me with a variety of problems for illustration."

The Hutcheson Bridge was completed in 1832. The masonry of the piers was laid at the level of seven feet below the bed of the Clyde, on a platform of timber, on piles eighteen feet in length. I found by

PLATE. VI.

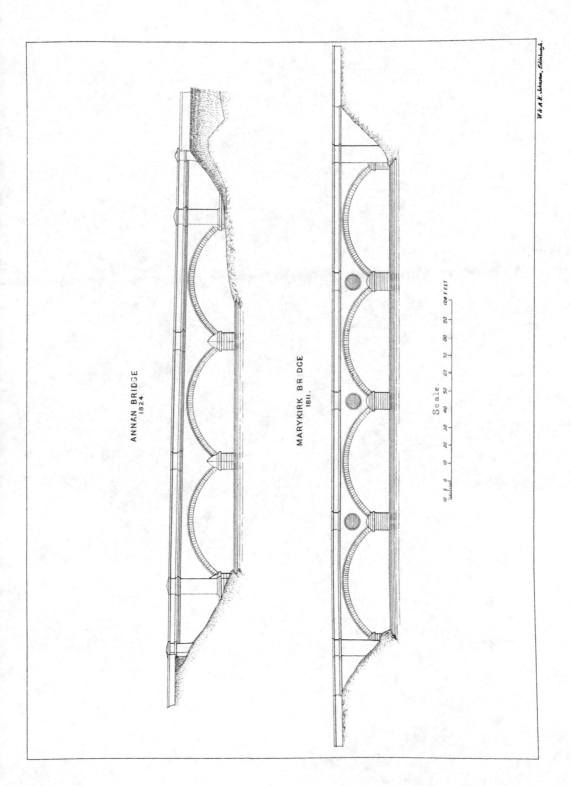

ANNAN BRIDGE
1824.

MARYKIRK BRIDGE
1811.

Scale.

10 5 0 10 20 30 40 50 60 70 80 90 100 FEET

W & A K Johnston, Edinburgh.

PLATE. VII.

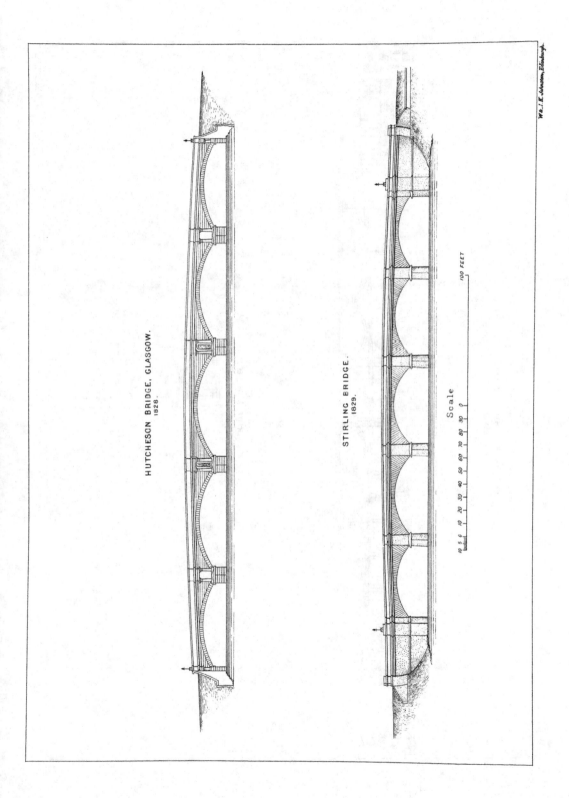

HUTCHESON BRIDGE, GLASGOW.
1828.

STIRLING BRIDGE.
1829.

Scale

10 5 0   10   20   30   40   50   60   70   80   90   100 FEET

Wᵐ. & E. Johnston, Edinburgh.

PLATE. VI.

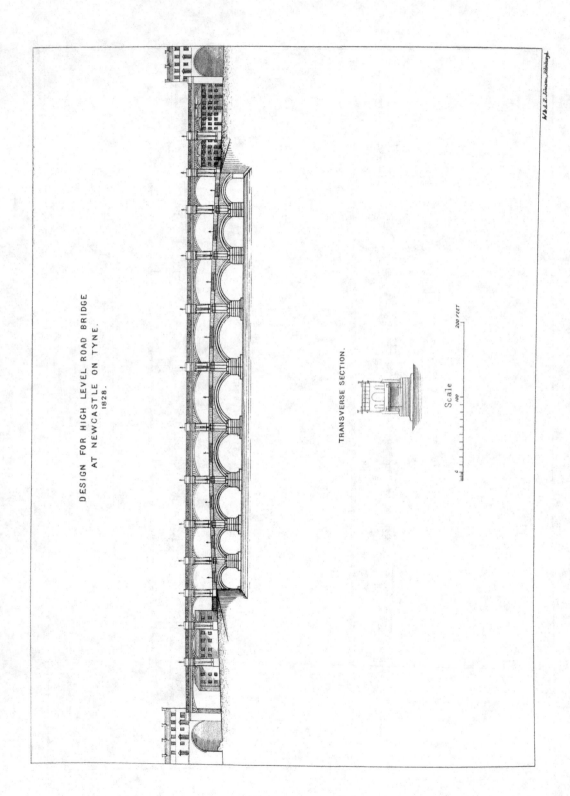

DESIGN FOR HIGH LEVEL ROAD BRIDGE
AT NEWCASTLE ON TYNE.
1828.

TRANSVERSE SECTION.

Scale

200 FEET

a section made in 1845, after a lapse of thirteen years, that the level of the river had been lowered, in consequence of the deepening of the river Clyde by the Navigation Trustees, no less than eleven feet, and even with that amount of scour the bridge was, and might long have remained, a safe structure. But immediately above its site there is a weir which dams up the Clyde and forms a lake, or almost still pool, in the river's bed for several miles. It was determined, in the interests of navigation, to take powers to remove the weir, and on its removal the bridge could no longer be pronounced safe; it was also resolved to take powers to replace the Hutcheson by the new Albert Bridge, designed by Messrs. Bell and Miller.

Mr. Stevenson has also left behind him some traces of originality of design in bridge-building.

In 1826 he gave a design to the Corporation of Newcastle for raising on the existing bridge another roadway, on a high level, to communicate with the higher parts of the town, as shown in Plate VIII., being the idea since so successfully carried out on a large scale by the late Mr. Robert Stephenson in his justly celebrated "high-level railway viaduct." Mr. Stevenson's design, as will be seen, consists of piers of masonry raised on the piers of the old bridge supporting a roadway of cast iron. The upper bridge being continued across the quays on either side of the river, and joining the roadways leading towards the south and north by easy gradients, avoided the circuitous and dangerous route of the old post road through Newcastle.

For timber bridges Mr. Stevenson also proposed, in

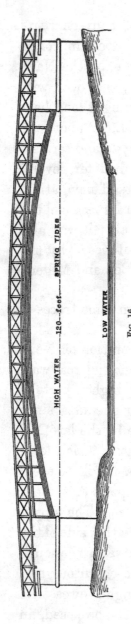

Fig. 16.

1831, a new form of arch of a beautiful and simple construction (Fig. 16), in which what may be called the " ring-courses " of the arch are formed of layers of thin planks bent into the circular form and stiffened by *kingpost pieces,* on which the level roadway rests. This form of bridge was afterwards very generally employed for railway bridges before the discovery had been made that for such works, structures of iron were, in the end, more economical than timber.

In 1820, he proposed to the Cramond District of Road Trustees, with a view mainly to lessening the cost of the work, a form of suspension bridge applicable to spans of moderate width, in which the roadway passes *above* the chains, and the necessity for tall piers is avoided. The suspension bridge over the Rhone at Geneva, and other bridges, have since been constructed on this principle.

In 1821 Mr. Stevenson wrote an article on Suspension Bridges for the *Edinburgh Philosophical Journal;* and as it contains a description of this new form of construction, as well as some historical information relative to bridges on the suspension principle, a few extracts from the paper may not be without interest :—

"*Winch Chain Bridge.*—The earliest bridges of suspension of which we have any account are those of China, said to be of great extent; Major Rennell also describes a bridge of this kind over the Sampoo in Hindostan, of about 600 feet in length. But the first chain bridge in our own country is believed to have been that of Winch Bridge over the river Tees, forming a communication between the counties of Durham and York. This bridge is noticed and an elevation of it given in the third volume of Hutchison's *Antiquities of Durham*, printed at Carlisle in 1794. As this volume is extremely scarce, owing to the greater part of the impression having been accidentally destroyed by fire, the writer of this article applied for a sight of it from the library of his friend, Mr. Isaac Cookson of Newcastle-upon-Tyne. The following account is given by Hutchison at p. 279 :—'The environs of the river (Tees) abound with the most picturesque and romantic scenes; beautiful falls of water, rocks and grotesque caverns. About two miles above Middleton, where the river falls in repeated cascades, a bridge suspended on iron chains is stretched from rock to rock over a chasm nearly sixty feet deep, for the passage of travellers, but particularly of miners; the bridge is seventy feet in length, and little more than two feet broad, with a hand-rail on one side, and planked in such a manner that the traveller experiences all the tremulous motion of the chain, and sees himself suspended over a roaring gulf, on an agitated and restless gangway, to which few strangers dare trust themselves.' We regret that we have not been able to learn the precise date of the erection of this bridge, but from good authority we have ascertained that it was erected about the year 1741.

"*American Bridges of Suspension.*—It appears from a treatise on Bridges by Mr. Thomas Pope, architect, of New York, published in that city in the year 1811, that eight chain bridges have been erected upon the catenarian principle, in different parts of America. It here deserves our particular notice, however, in any claim for priority of invention with our transatlantic friends, that the chain bridge over the Tees was known in America, as Pope quotes Hutchison's vol. iii., and gives a description of Winch Bridge. It further appears from this work that a patent was granted by the American Government for the erection of bridges of suspension in the year 1808. Our American author also describes a bridge of this construction, which seems to have been erected about the year 1809, over the river Merrimack in the State of Massachusetts, consisting of a catenarian arch of 244 feet span. The roadway of this bridge is suspended between two abutments or towers of masonry, thirty seven feet in height, on which piers of carpentry are erected which are thirty five feet in height. Over these ten chains are suspended, each measuring 516 feet in length, their ends being sunk into deep pits on both sides of the river, where they are secured by large stones. The bridge over the Merrimack has two carriage-ways, each of fifteen feet in breadth. It is also described as having three chains which range along the sides, and four in the middle, or between the two roadways. The whole expense of this American work is estimated to have been 20,000 dollars.

"*Proposed Bridge at Runcorn.*—Perhaps the most precarious and difficult problem ever presented to the consideration of the British engineer was the suggestion of some highly patriotic gentlemen of Liverpool, for constructing a bridge over the estuary of the Mersey at Runcorn Gap, about twenty miles from Liverpool. The specifications for this work provided that the span of the bridge should measure at least 1000 feet, and that its height above the surface of the water should not be less than sixty feet, so as to admit of the free navigation of this great commercial river. The idea of a bridge at Runcorn, we believe, was first conceived about the year 1813, when the demand for labour was extremely low, and a vast number of the working classes of Lancashire were thrown out of employment. A variety of designs for this bridge were procured by a select committee of the gentlemen who took an interest in this great undertaking. The plan most approved of, however, was the design of a bridge of suspension; and Mr. Telford the engineer, and Captain Brown of the Royal Navy, are understood pretty nearly to have concurred in opinion as to the practicability of such a work. Mr. Telford has reported fully on the subject, and has estimated the expense of his design at from £63,000 to £85,000, according to different modes of execution. Though as yet little advancement has been made in carrying this enterprising design into execution, yet the novelty and magnitude of an arch of 1000 feet span is a subject of so much interest that we have thought it proper in this place to mention these circumstances.

"*Menai Chain Bridge.*—The Straits of Menai, which separate the island of Anglesea from Caernarvonshire, have long formed a troublesome obstruction upon the great road from London to Dublin by Holyhead, by which the troublesome ferry of Bangor might be avoided. Many plans for the execution of this undertaking have also been agitated, chiefly in cast iron, including a range of estimate from about £128,000 to £268,000; but that which is now acted upon is a bridge of suspension upon the catenarian principle, the extent of which between the piers or points of suspension is to be 560 feet, the estimate for which is only about £70,000. This by many has been considered a work of great uncertainty; but the Union Bridge on this plan has already been executed on the Tweed, to the extent of 361 feet."

Mr. Stevenson then goes on to mention several wire and chain bridges erected in Scotland, and gives the following description of his design for Cramond Bridge:—

" Fig. 17 is a section and plan designed for crossing the river Almond on the great north road between Edinburgh and Queensferry. The extent of the span between the points of suspension is laid down at 150 feet. The chief circumstances which particu-

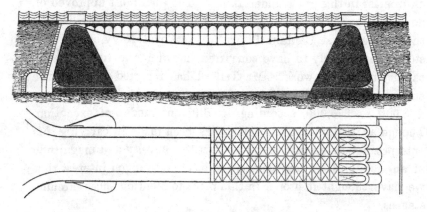

Fig. 17.

larise this design are a mode of fixing the chains to the abutments of suspension on each side of the river, by which the main chains can be distributed equally under the roadway. The main chains are likewise made to collapse or turn round the abutments of masonry, as will be seen from the section, in which the parts of the work are so contrived that access can be had to the chains by an arched way on each side. In this design the two ends of the chains are formed into great *nails* or bolts, with countersunk or conical heads made to fit into corresponding hollow tubes of cast iron built into the masonry of the abutments.

" From this description the reader will readily form an idea of the simplicity and effect of this mode of fixing the chains, being such, also, that any particular chain may be withdrawn and replaced without deranging the fabric of the bridge. The roadway, instead of being *suspended* from the main chains, is made up to the proper level upon the chains by a framework of cast iron, prepared for the reception of a stratum of broken stones for the road.

" The making up of the roadway of this bridge, however, and the enlarged angle of its suspension, may be considered as limiting the span or extent of bridges of this construction to about 200 feet. The structure represented by Fig. 17 appears to possess many advantages for bridges of that modified extent, and the manner of fixing the chains is applicable to all bridges of suspension; it is likewise new, so far as we know."

In the close of his paper Mr. Stevenson says :—

" To what extent suspension bridges may be carried is very uncertain, and he who has the temerity to advance sceptical or circumscribed views on this subject would do well to reflect upon the history of the steam-engine. When the Marquis of Worcester first proposed, by the boiling of water, to produce an effective force, no one could have conceived the incalculable advantages which have since followed its improvement by our illustrious countryman, Watt."

A prophetic announcement, which has had its full realisation in the Suspension *Railway* Bridge of 821 feet span at Niagara Falls, and in the still bolder design now in execution for connecting New York and Brooklyn by a steel wire suspension bridge, having a clear opening between the piers of no less than 1600 feet.

CHAPTER XI.

WOLF ROCK LIGHTHOUSE.

ABOUT the year 1812, Mr. Stevenson having, as adviser of the Commissioners of Northern Lighthouses, attained the position of being the most eminent Lighthouse Engineer of his day, was requested by the Admiralty to report on the practicability of erecting a lighthouse on the Wolf Rock, lying about eight miles off the Land's End in Cornwall.

I give, from Mr. Stevenson's "Journal," the following curious account of the first visit he made to the rock; and it may perhaps be as well to say that all quotations made from what I have called his "Journal" are records of what he roughly noted down at the time in the form of a Diary, and are on that account perhaps all the more interesting, at least to non-professional readers.

"14th Sept. 1813.—Waited upon Sir Robert Calder, Admiral of the port of Plymouth, on the 13th, in consequence of letters from Lord Melville relative to a vessel to carry me to the Wolf Rock.

"The Admiral accordingly appointed the Orestes,' Captain Smith, to proceed with me to the Wolf, and after landing me there, and having made my observations, Captain Smith was directed to land me at any port most convenient for me, according to the state of the weather. Captain Smith, in consequence of this order, and to suit my convenience, got the 'Orestes' in readiness two days sooner than he otherwise intended, and I embarked on the 14th at 2 P.M. agreeably to appointment.

" The Captain took me by the hand and welcomed me on board His Majesty's ship, and introduced me to his first lieutenant, Mr. Fallick. He then proceeded to give orders for casting off, which was done in an instant after the word was given. The 'Orestes' is properly a gun brig, but rigged as a ship, has 28 guns and 100 men. Kept plying to windward, and in the evening had the Eddystone light in view, still upon our lee quarter, distant eight or ten miles.

" *15th.*—Kept working along the shore all day, and at 7 P.M. a pilot from Mousehole by Penzance came on board. Upon consulting the pilot, he recommended that the ship should be brought to an anchor in Mounts Bay, or rather Newland Road, all night, as it would answer no good purpose to go round the land so soon after a fresh gale of wind, with the view of landing on the Wolf, which he represented as being only practicable in the finest of summer weather. This was poor heartening. The Captain submitted to me whether it were not more advisable to come to an anchor, in which, with all submission to him, I consented. The ship, accordingly, was brought to an anchor in twelve fathoms, clean sand.

" On board of the 'Orestes' two of the people were punished, —one for threatening to *knock down* the serjeant of marines, while on duty, received three dozen; another who offered an insult to a lieutenant, received one dozen.

" I was sitting below, the time this was going forward, when all hands were piped on deck, and the Captain began to read the Articles of War. He had previously said to me that two men were in irons, whom he meant to punish and liberate. I went upon deck to learn the cause of all being so quiet, and discovering what was intended, I went below and waited in great suspense till the men began to call out for mercy. I took the liberty of sending a note

Y

to the Captain—the circumstances were so painful to me—to see if he could remit any part of the punishment, to which I afterwards understood he had listened, as he did not give them so many lashes as was intended. Captain Smith had by no means the character of a severe commander, as I understood from some of the officers he had been two years in the ship, and had only punished twice.

"About 9 P.M., while the Captain and myself were at supper, we heard a conversation between the pilot and Mr. Fallick, the first lieutenant, about a vessel being on fire. The former was of opinion that it was a pilchard boat, the crew of which were roasting pilchards, while Mr. Fallick insisted that it was a vessel on fire. In a short time the vessel or boat appeared to be in flames, and with all sail set she approached the 'Orestes.' On shore the people of Penzance and Mousehole were afraid of the 'Orestes' taking fire and discharging a broadside upon the town. In the meantime the vessel on fire approached the 'Orestes' so directly that Captain Smith gave orders to veer out all the cable, stand by to cut or bend on more rope, according to circumstances.

"The weather became moderate, and we had little or no wind, and the vessel on fire (which turned out to be a sloop of 80 or 90 tons, bound for St. Sebastian with bottled porter and bale goods) passed ahead of the 'Orestes' about half a cable's length. Her hull was then completely on fire, but the rigging and sails had not then caught fire, and she kept an undeviating course till she grounded on the shore.

"Captain Smith then despatched officers and men in three boats to endeavour to save as much as possible, but a report having gone abroad that she had gunpowder on board no person ventured near the vessel on fire till it was too late to be of any service, and in the morning when Captain Smith and I went on shore nothing remained but the keel and a few of the 'futtocks' half burned, and the mast over by the deck, the lower part having been consumed by the flames. The vessel was just getting under weigh when the accident occurred, through the carelessness of a boy, who set a lighted candle into a crate of straw in which bottles were packed. The crew soon afterwards appear to have carelessly deserted the vessel and landed at Mounts Bay, three miles from Mousehole, and appear not to have been very active in doing what was in their power. The loss of ship and cargo was estimated at £14,000.

"16*th.*—Got under weigh at 6 P.M., and left Mousehole Bay with an intention to go round the land; but the weather fell calm, and after shutting in the Lizard lights came to an anchor in Mounts Bay till next morning. The Lizard lights appeared to very great advantage.

"17*th.*—Got under weigh at 6 A.M., wind shifting from south-west to east with a fine breeze, and at 11 A.M. got up with the Wolf Rock. At 12 noon two boats were manned—one commanded by a midshipman, and the other by Lieutenant Fallick, into which I went, and after pulling round and round the rock with both boats, sounding all the while, we made preparations for landing. Mr. Fallick arranged his boat's crew, and let go a grapling over the stern, then veered away upon this stern rope watching a smooth, and when the boat was near enough the young man (the same who had two days before got one dozen of lashes) appointed to land with a bow rope to make fast, leaped upon the rock, and upon these two ropes the boat was hauled off and on with great ease and facility. In this manner Lieutenant Fallick landed next, then I landed, but not without much difficulty, and watching an opportunity to get on the rock with a smooth between the seas.

"Upon leaving the ship, about a quarter of a mile from the rock, I began to sound, and at from two to three cables' length off the rock have 41, 40, and 38 fathoms water, with shell sand of a fair colour. At about one cable's length have 13 fathoms, same bottom. Within this distance have 10, 8, 5, 3½; and 2 fathoms, chiefly rocky bottom.

" The rock is steep in all directions; the south-west if anything draws to a point with rather less water near it than in other directions.

"At low water of a neap tide the rock appeared to be about twelve or fourteen feet in perpendicular height above the surface of the water. Its surface is very irregular, jutting up in masses of from six to ten feet in height. These inequalities all presented marked and angular outlines, terminating in well-defined points and edges. The central part of the rock is formed pretty much into a hollow, where there have been some quarrying operations in fixing the beacon which was erected upon it. The margin of the rock is upon the whole pretty regular, as it appears jutting out of the water. On the eastern side it is not so regularly formed at the water's edge as on the western side. It slopes outwards, and seems to form a large stool in every direction. At some places there are guts or slips in the rock, but none of these are large enough to be useful for a boat landing at. The best and perhaps the only landing place is at the north-east side, where the rock is most precipitous.

"Taking the dimensions in the largest directions with the lead-line, in fathoms, it measured twenty-two fathoms in a north-east and south-west direction, and sixteen fathoms in a north-west and south-east direction.

" Upon the surface in the middle, at the hollow place, I found a hole of six inches in depth, and about nine inches square, and connected with it, at six feet distant, three holes for bats, which I presume to have been the step of the beacon, and the iron bats were still to be seen which had been used as guys. This fragile affair appears to have wanted base and every requisite suited to such an exposed situation and important purpose, and accordingly the beacon, with a wolf of metallic work, erected by a Lieutenant Smith, who erected the Longships Lighthouse, is said not to have remained longer than a few days, and was carried away in the first storm.

" Besides these holes and bats, which last seem not to have exceeded 1½ inch iron in strength, I found several eye bolts in different parts of the rock, particularly at the landing place, which had been put in to make fast boats, etc., while the beacon was being erected.

" The surface of the rock is extremely rugged, and running in every direction into sharp angular points. The rock seems to run in beds from an inch to a foot in thickness. It has much the appearance of limestone, but upon a narrow inspection it turns out to be porphyry. It is covered with the barnacle, many limpets of a very large size—say two inches diameter,—and mussels. These were the only animal productions that were found upon it. Of the marine *fuci* there were two or three varieties.

" That it would be practicable to erect a building upon this rock I have no doubt, but from its shape and figure, and the great depth of water in all directions round it, together with the smallness of its dimensions, it would be a work of great difficulty; and be attended with much expense and great hazard.

" I am therefore of opinion that it might cost from £80,000 to £90,000 to erect a lighthouse at the Wolf, with all the requisite buildings and appointments, like the Bell Rock Lighthouse.

" In a conversation on this subject with Lieutenant Smith in 1806 (who had erected the beacon on the Wolf), he pronounced it as an impracticable work. But his opinion, from the work he had performed at the Longships, and other circumstances, made very little impression upon my mind, at the time, in regard to the Bell Rock, and since seeing the Wolf Rock I think his arguments were ill founded, and I am perfectly decided in opinion that the work is a practicable one.

" The wind being nearly easterly, and consequently unfavourable for returning with dispatch to Plymouth, the captain gravely proposed that we should stand towards 'the Bay' for a few days, when it might shift. Not being fully aware of what was meant by the Bay, I put the question, when to my surprise he meant the Bay of Biscay, and said we should see St. Sebastian, which had just fallen; but to this I replied, that I should much rather be landed at the Land's End. He was constantly on the outlook for prizes, and as I came not to fight I wanted much to be on shore, that I might pursue my way to Bath, where I knew Mr. Rae, the Sheriff of Edinburgh, would be waiting my return to proceed upon the visit to the Prisons on our return to Scotland.

"The ship was therefore directed to steer for the Land's End, and the pilot took the ship within the Longships Lighthouse, and he and I landed at Sennan on the same evening.

"Having procured horses for myself and luggage, I set off immediately for Penzance, which I reached about 10 o'clock at night, the 17th September, much pleased with my trip upon the whole.

"18*th*.—Leave Penzance, and reach Falmouth by the fly.

"19*th*.—Leave Falmouth, and that same night, or early next morning, reach Exeter.

"20*th*.—At 6 A.M. leave Exeter, and 8 P.M. reach Bath.

"From Plymouth to the Wolf, and returning to Bath, only eight days."

Mr. Stevenson at a subsequent date made another visit to the Wolf, accompanied by an assistant, when a careful survey was made, followed by a well-considered design, which is shown in Plate IX., and is described by him as follows :—

"Plate IX. is the section of a design formed by the revolution of the parabola round the axis of a building, as its asymptote, whose base measures fifty-six feet in diameter, and parallel at the top of the solid is thirty-six feet; and height to the entrance door, thirty-five feet. The contents of this figure between these

PLATE. IX.

Scale of Feet to Fig.ˢ 1. 2. 3. 4. 5

10 5 0   10   20   30   40   50

DESIGN FOR WOLF ROCK LIGHT HOUSE.

parallels is calculated at 45,000 cubic feet; but the whole of the masonry of the design is estimated at 70,624 cubic feet. Its general features may be stated as similar to those of the Eddystone and Bell Rock Lighthouses, the parts being only enlarged, and the parabolic instead of the logarithmic curve adopted for its outline. In this design, the parabolic curve is continued from the basement to the copestone of the light room, exclusively of the projection for the cornice and balcony. The masonry is intended to be 120 feet in height, estimating from the medium level of the sea, of which the solid, or from the foundation to the entrance door, forms thirty-five feet, the staircase twenty-five feet, and the remaining sixty feet of its height is occupied with six apartments, and the walls of the light room. In the staircase a recess is formed for containing the machinery for raising the stores to the height of the entrance door; here a small hole is perforated through the building for the admission of the purchase chain. The thickness of the walls immediately above the solid is twelve feet; at the top of the stone staircase they are eight feet, and where the walls are thinnest, immediately under the cornice, they measure two feet. A drop hole formed in the courses of the staircase and solid, provides for the range of the weight of a revolving light. The ascent to this building, as at the Bell Rock, is intended to be by an exterior stair or ladder of brass, and the interior communication between the several apartments by means of flights of circular oaken steps."

The only estimate Mr. Stevenson ever made of the work was that already stated in his Journal, at a cost of £80,000 to £90,000 for the tower and requisite dwellings for the lightkeepers and crew of attending vessel ashore.

Mr. Stevenson's original visit was, as we have seen, made in 1813, and in 1870, after a lapse of fifty-seven years, the present tower on the Wolf Rock, the joint work of the late Mr. James Walker and of Mr. James N. Douglass, was successfully accomplished under the auspices of the Trinity House. The cost of the tower, exclusively of the shore establishment, which it was unnecessary to provide, was £62,726, being not very different from the estimate of Mr. Stevenson (from £80,000 to £90,000), which included a shore establishment.

CHAPTER XII.

CARR ROCK BEACON.

1810—1821.

THE Carr Rock is a tide-covered reef extending about $1\frac{3}{4}$ mile from the shore of Fifeness, and forming a *turning point* in the navigation of the northern-bound shipping of the Firth of Forth, and on Mr. Stevenson's recommendation the Commissioners of Northern Lighthouses resolved to erect a beacon of masonry to mark the danger.

It may seem to be unnecessary, after describing the Bell Rock Lighthouse, to notice so apparently small a work as this ; but in such matters it is unsafe to generalise ; each case must be considered on its own merits, and great difficulties were encountered in accomplishing the work. The formation of the Carr Rock rendered it impracticable to secure a base for a building of greater diameter than eighteen feet, and as part of that base had to be founded under the level of the lowest tides by coffer-dams which were removed and taken ashore after each tide's work, even the Engineer of the Bell Rock Light-house found all his resources taxed to a considerable extent, and he was in the end foiled in carrying out his design for the building. But irrespectively of these

z

physical difficulties, the Carr Rock is a work of great
interest to the lighthouse engineer, inasmuch as Mr.
Stevenson at that early date conceived the idea of calling
to his aid the power given by the rise of tide on the
building to move a train of clock work to sound a warning
bell; and again, when the destruction of the upper
portion of his beacon by the sea obliged him to relinquish
this plan, unwilling to be beaten, he suggested that the
same tidal action might be made to sound a whistle; and
failing that, he proposed to exhibit a phosphorescent light
from the top of the building. All of these ideas suggested
by Mr. Stevenson's inventive mind have been from time
to time revived by modern inventors.

The original design of the Carr Rock Beacon was made
in 1810, and the work was commenced in 1813. After
portions of the masonry had repeatedly been carried
away by the sea, the original design for surmounting the
building by a bell to be rung by the rise and fall of the
tide was abandoned, and the beacon was completed in
1821, by raising an iron structure, as shown in Plate x.
Fig. 2, on the foundation that had escaped the fury of the
sea, and that structure is still in perfect preservation.
So great, indeed, was the difficulty that Mr. Stevenson, in
1818, contemplated using blocks of cast iron instead of
stone to insure greater specific gravity—a proposal which
is believed to have been then made for the first time.

The following is Mr. Stevenson's own description of
this interesting work :—

"The form and construction of the Carr Rock Beacon,
as originally designed and ultimately executed, will be

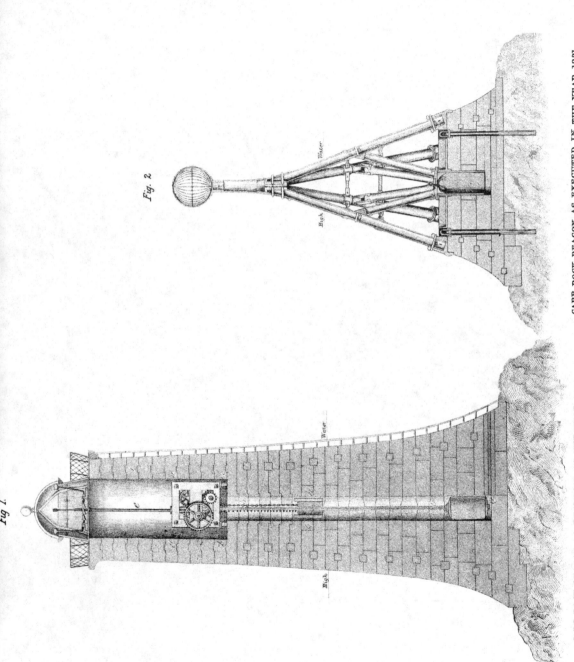

PLATE X.

Fig. 1.

Fig. 2.

CARR ROCK BEACON AS DESIGNED IN THE YEAR 1810

CARR ROCK BEACON AS EXECUTED IN THE YEAR 1821

Scale of Feet
0 1 2 3 4 5 6 7 8 9 10   15   20

better understood by referring to Plate x. The motion originally intended to be given to the bell-apparatus, or tide machine, Fig. 1, was to be effected by admitting the sea through a small aperture of three inches in diameter, perforated in the solid masonry, communicating with a cylindrical chamber in the centre of the building, measuring two feet in diameter, in which a float or metallic air tank was to rise and fall with the tide. During the period of flood tide, the air vessel, in its elevation by the pressure of the water, was to give motion to machinery for tolling the bell and winding up a weight, which last, in its descent, during ebb tide, was to continue the motion of the machine, until the flood tide again returned to perform the joint operation of tolling the bell and raising the weight. A working model of a machine upon this principle having been constructed, it was kept in motion for a period equal to several months ; this was effected by water run through a succession of tanks raised by a pump from the lower one to the higher, thus producing the effect of flood and ebb tides. The time during which this apparatus was in action having been ascertained by an index, a constant attendance upon the machine during this protracted experiment became unnecessary.

" The upper termination of the beacon, in its present form, as shown in Fig. 2, does not admit of the application of the tide machine with the bell apparatus. Experiments as applicable to this have, however, been tried with a wind instrument, to be sounded by the pressure of the sea water, but it has not succeeded to the extent that seems necessary for a purpose of this kind. We have, indeed, thought that the application of pressure as a power, communicated by the waters of the ocean, in mechanical operations, might be carried to almost any extent by simply providing a chamber or dock large enough for the reception of a float or vessel, of dimensions equivalent to the force required. This description of machinery is more particularly applicable in situations where the tides have a great rise, as in the Solway Firth, Bristol Channel, and other parts of the British seas ; and at St. Malo on the coast of France.

"A beacon of any form, unprovided with a light, must always be considered an imperfect landmark, and therefore various modes have been contemplated for more completely pointing out the position of the Carr Rock. It has been proposed that phosphoric lights should be exhibited from the top of the building. This object, however, would be more certainly accomplished by the erection of leading lights upon the island of May and mainland of Fife. But these, with other plans, which have been under the writer's consideration, would necessarily be attended with a great additional expense, which, in the present instance, it is not thought advisable to incur."

CHAPTER XIII.

CRANES.

IT appears that Mr. Stevenson was much perplexed as to what sort of cranes he should use in building the Bell Rock Lighthouse. His difficulties were twofold :—

First, In consequence of the dovetailed form of the stones he required a crane that would drop them as nearly as possible on the beds on which they were permanently to rest.

Second, Supposing he devised a *guy crane* that overcame this difficulty, what was to be done as the building rose in height, and the guys became too nearly perpendicular to admit of such a crane being used ?

In his private notes Mr. Stevenson regrets that he could get no advice from anybody he consulted, all of whom recommended him to employ common sheer poles, such as had been used by Smeaton at the Eddystone ; and he adds, "I may say, morning, noon, and night, these difficulties have haunted me." But thrown back on his own resources, and appreciating the difficulty as no one else could so well do, he found, as is often the case, that he was his own best counsellor, and he succeeded in solving the problem that had given him so much concern, by inventing what he called the "moveable beam

crane," and also the " balance crane," which are shown in
Plate XI. The former, as modified to suit particular
cases, is now in universal use for building purposes, and
the latter has been employed in rearing most of our Rock
Lighthouses, so that I think professional readers will
not object to my giving Mr. Stevenson's description
of these cranes, as designed by him at the beginning of
this century. He says :—

" In cranes of the common construction the beam is a
fixture, and is placed at right angles to the upright shaft :
but in the machine represented in the Plate (Fig. 1),
its attachment is at the lower extremity of the crane,
where it is moveable up and down upon a journal or
bolt. This crane is therefore termed a moveable beam
crane. The moveable property of the beam, in so far as
the writer knows, is new, and possesses the advantage of
laying any stone within its range perpendicularly on its
site. This, from the dovetailed form of the stones at the
Bell Rock, rendered it particularly fitted for this work, to
which a crane of the ordinary construction could hardly
be said to be applicable. At the Eddystone Lighthouse
this operation was performed by means of triangular
sheers ; but, from the greater extent of the Bell Rock
works, and their greater depth in the water, such means
must have rendered the process of building extremely
tedious. These cranes were necessarily immersed at
high water, and were retained in their places by four
guys fixed at the top of the upright shaft, and the moveable jib or beam being lowered down, was secured to an
eyebolt batted into the rock."

PLATE. XI.

Fig. 2

Fig. 1

Scale of Feet.

MOVEABLE JIB AND BALANCE CRANES.

" The ' balance crane' (Fig. 2) was constructed on a
new principle for building the upper part of the Bell Rock
Lighthouse, when the guy ropes of the moveable beam
crane became ' too taut,' as sailors express it, or were
too near the perpendicular, thereby rendering the beam
cranes unstable. To remedy this, the balance crane was
so arranged as to be kept in equilibrium by a back weight
of cast iron, so adapted as to counteract the varying load
upon the working arm or beam. The elevation here
represented is the same in principle with that used at the
Bell Rock, but differs somewhat in form, agreeably to
improvements made in order to adapt it to the erection
of the Carr Rock Beacon. The upright central column is
a tube of cast iron put together in convenient lengths
with flush joints, after the manner of spigot and faucet,
fitted by turning and boring. The centre column of this
machine might have been carried to any suitable or convenient height, by adding length to length, as the building
advanced, without once moving the foot on which it rested,
but at the Bell Rock not more than three lengths of from
six to nine feet were generally in use. A malleable iron
cross head was stepped into the void of the central shaft
or column when the body of the crane was to be elevated.
This operation was accomplished simply by hooking the
main 'purchase' and 'traveller' chains into the eyes of
the crosshead, when the machinery of the crane was
employed with great facility as a locomotive power for
lifting itself as each new length of central column was
added. The weight of this crane as used at the Carr
Rock did not exceed two tons."

CHAPTER XIV.

MR. STEVENSON was ever an intelligent and anxious observer of the habits and industry of the people of those remote and isolated parts of the country which he so often visited. He was specially interested in the fisheries from which they mainly derive their support, as testified by frequent allusions to them in his journals and notes.

The following notice regarding the state of the Scottish fisheries, made in 1819, to the editor of the *Edinburgh Philosophical Journal*,[1] will be read with interest :—

"Having been for many years conversant with the navigation of the Scottish seas, I have, prior to the war with Holland, seen fleets of Dutch 'busses' engaged in the herring fishery off the northern parts of our coast. For a long time past, however, those industrious fishermen had not ventured to approach these shores ; and they are now only beginning to reappear.

"In the early part of August last, while sailing along the shores of Kincardineshire, about ten miles off Dunnottar Castle, the watch upon deck, at midnight, called out 'Lights ahead.' Upon a nearer approach these lights were found to belong to a small fleet of Dutch fishermen

[1] Vol. ii. p. 129.

employed in the deep sea fishing, each vessel having a lantern at her mast head. What success these plodding people had met with our crew had no opportunity of inquiring; but upon arriving the next morning at Fraserburgh,—the great fishing station on the coast of Aberdeen —we found that about 120 boats, containing five men each, had commenced the fishing season here six weeks before, and had that night caught no fewer than about 1500 barrels of herrings, which in a general way, when there is a demand for fish, may be valued at £1 sterling per barrel to the fishermen, and may be regarded as adding to the wealth of the country perhaps not less than £3000. In coasting along between Fraserburgh and the Orkney Islands, another fleet of Dutch fishermen was seen at a distance. The harbour and bay of Wick were crowded with fishing boats and busses of all descriptions, collected from the Firth of Forth and southward even as far as Yarmouth and Lowestoft. The Caithness fishing was said to have been pretty successful, though not equal to what it has been in former years.

"In the Orkney and Shetland Islands one would naturally look for extensive fishing establishments, both in herrings, and what are termed white fish (cod, ling, and tusk); but it is a curious fact, that while the Dutch have long come from their own coast to these islands to fish herrings, it is only within a very few years that the people of Orkney, chiefly by the spirited and praiseworthy exertions of Samuel Laing, Esq., have given any attention to this important source of wealth. It has long been a practice with the great fishmongers of London to send

2 A

their *welled* smacks to fish for cod, and to purchase lob-
sters, around the Orkney Islands ; and both are carried
alive to the London market. This trade has done much
good to these islands, and has brought a great deal of
money to them ; but still it is of a more circumscribed
nature, and is less calculated to swell the national wealth,
than the herring and white fishery in general.

" Hitherto the industry of the Orcadians has been
chiefly directed to farming pursuits ; while the Shetlanders
have been almost exclusively occupied in the cod, ling,
and tusk fishing. It is doubtful, indeed, if, up to this
period, there be a single boat belonging to the Shetland
Isles which is completely equipped for the herring fishery.
But on reaching Shetland another fleet of Dutch doggers
was seen collecting in numbers off these islands—a coast
which is considered a rich harvest in Holland.

" So systematically do the Dutch pursue the fishing
business upon our coasts, that their fleet of busses is
accompanied by an hospital ship. This vessel we now
found at anchor in Lerwick roads, and were informed
that she paid weekly visits to the fleet, to supply medi-
cines, and to receive any of the people falling sick, or
meeting with any accident.

" Though Shetland is certainly not so much an agricultural country as Orkney, yet it may be hoped that the
encouragement judiciously held out by the Highland
Society, for the production of green crops in Shetland,
may eventually have the effect of teaching these insular
farmers the practicability of providing fodder for their
cattle in the spring of the year. This has long been a
great desideratum. The command of a month or six

weeks' fodder would enable the proprietors of that country to stock many of their fine verdant isles with cattle, and to employ their hardy tenantry more exclusively in the different branches of the fishery.

"It is well known, that, next to the Newfoundland Banks, those of Shetland are the most productive in ling, cod, tusk, and other white fish; and by the recent discovery of a bank, trending many leagues to the southwestward, the British merchants have made a vast accession to their fishing grounds. The fishermen who reside in the small picturesque bay of Scalloway, and in some of the other bays and voes on the western side of the mainland of Shetland, have pursued with much success the fishing upon this new bank, which I humbly presume to term the REGENT FISHING BANK—a name at once calculated to mark the period of its discovery, and pay a proper compliment to the Prince. Here small sloops, of from fifteen to twenty-five tons burden, and manned with eight persons, have been employed In the beginning of August they had this summer fished for twelve weeks, generally returning home with their fish once a week. On an average, these vessels had caught 1000 fine cod fish a week, of which about 600 in a dried state go to the ton, and these they would have gladly sold at about £15 per ton. So numerous are the fish upon the Regent Fishing Bank, that a French vessel, belonging, it is believed, to St. Malo, had sailed with her second cargo of fish this season; and though the fishermen did not mention this under any apprehension, as though there were danger of the fish becoming scarce, yet they seemed to regret the circumstance, on account of their market being thus preoccupied.

"Here, and at Orkney, we had the pleasure to see many ships arriving from the whale fishing, and parting with a certain proportion of their crews. To such an extent, indeed, are the crews of the whalers made up from these islands, that it is calculated that not less than £15,000 in cash are annually brought into the islands by this means. With propriety, therefore, may the whale fishery be regarded as one of the most productive sources of national wealth connected with the British Fisheries.

"From the Orkney and Shetland Islands our course was directed to the westward. A considerable salmon fishing seems to be carried on in the mouths of the rivers of Lord Reay's Country in Sutherlandshire: the fish are carried from this to Aberdeen, and thence in regular trading smacks to London. We heard little more of any kind of fishing till we reached the Harris Isles. There, and throughout the numerous lochs and fishing stations on the mainland, in the districts of Gairloch, Applecross, Lochalsh, Glenelg, Moidart, Knoidart, Ardnamurchan, Mull, Lorn, and Kintyre, we understood that there was a general lamentation for the disappearance of herrings, which in former times used to crowd into lochs which they seem now to have in some measure deserted. This the fishermen suppose to be owing to the *Schools* being broken and divided about the Shetland and Orkney Islands; and they remark, that, by some unaccountable change in the habits of the fish, the greatest number now take the east coast of Great Britain. This is the more to be regretted, that in Skye, the Lewis, Harris, and Uist Islands, the inhabitants have of late years turned their attention much to the fishing. Indeed, this has followed as a matter of necessity, from the general practice of converting the numerous small arable farms, which were perhaps neither very useful to the tenants nor profitable to the laird, into great sheep walks; so that the inhabitants are now more generally assembled upon the coast. The large sums expended in the construction of the Caledonian Canal have, either directly or indirectly, become a source of wealth to these people : they have been enabled to furnish themselves with boats and fishing tackle, and for one fishing boat which was formerly seen in the Hebrides only twenty years ago, it may be safely affirmed that ten are to be met with now. If the same spirit shall continue to be manifested, in spite of all the objections which have been urged against the salt laws, and the depopulating effects of emigration, the British Fisheries in these islands, and along this coast, with a little encouragement, will be wonderfully extended, and we shall ere long see the Highlands and Islands of Scotland in that state to which they are peculiarly adapted, and in which alone their continued prosperity is to be looked for, viz., when their valleys, muirs, and mountains are covered with flocks, and the people are found in small villages on the shores."

The following history of the origin of the Shetland herring fishery, communicated to *Blackwood's Magazine* in 1821, is, I think, worthy of being recorded :—

" Few people, on examining the map of Scotland, would believe that the herring fishing has only within these few years been begun in Orkney, while the natives are almost strangers to the fishing of cod and ling.

"On the other hand, it is no less extraordinary that although the cod and ling fishery has been carried to so great an extent in Shetland as to enable them to export many cargoes to the Catholic countries on the Continent, not a herring net has been spread by the natives of Shetland till the present year (1821), when Mr. Mowat of Gardie, and a few other spirited proprietors of these islands, formed themselves into an association, and subscribed the necessary funds for purchasing boats and nets, to encourage the natives to follow the industrious example of the Dutch.

"The immediate management of this experimental fishery was undertaken in the most patriotic and disinterested manner by Mr. Duncan, the Sheriff-Substitute of Shetland. Having procured three boats, he afterwards visited Orkney, to ascertain the mode of conducting the business there, and having also got fishermen from the south, this little adventure commenced. Its nets were first wetted in the month of July, and it is believed its labours were concluded in the month of September, after obtaining what is considered pretty good success, having caught as follows, viz. :—

The 'Experiment,' 6-manned boat,	.	.	$212\frac{1}{2}$ crans.
The 'Hope,'	5 "	. .	$119\frac{3}{4}$ "
The 'Nancy,'	4 "	. .	80 "
			$412\frac{1}{4}$ "

"The great object which the Shetland gentlemen have in view, in this infant establishment, is to give employment to their fishermen in the herring trade, after the cod and ling season is over, and by this means to enable them to partake of those bounties and encouragements so properly bestowed by Government on the fisheries; and thus abstract the attention of the lower orders of these islands from an illicit traffic in foreign spirits, tea, and tobacco, which has greatly increased of late years.

"The profit of the herring fishing at its commencement has, however, afforded more encouragement than could have been expected; for, besides paying the men a liberal allowance for their labour, a small sum has been applied towards defraying the expense of the boats and nets. But what is of far more consequence to this patriotic association is the spirit of enterprise which it is likely to create by bringing forward a number of additional boats in the way of private adventure, which must be attended with the best advantage to the Shetland Islands."

THE SYMPIESOMETER.

Again, in 1820, Mr. Stevenson took occasion to express his solicitude for the welfare of the fishermen in the following note, suggesting the means whereby they might sometimes avoid a coming storm—a suggestion which is now to some extent carried out by the Board of Trade's establishment of marine barometers at many of our fishing stations :—

"Mr. Stevenson informs us," says the editor of the *Edinburgh Philosophical Journal*[1] for 1820, "that having occasion, in the beginning of September last, to visit the Isle of Man, he beheld the interesting spectacle of about 300 large fishing boats, each from fifteen to twenty tons

[1] Vol. ii. p. 196.

burden, leaving their various harbours at that island in an apparently fine afternoon, and standing directly out to sea with the intention of prosecuting the fishery under night. He at the same time remarked that both the common marine barometer, and Adie's sympiesometer, which were in the cabin of his vessel, indicated an approaching change of weather, the mercury falling to 29·5 inches. It became painful, therefore, to witness the scene,—more than a thousand industrious fishermen, lulled to security by the fineness of the day, scattering their little barks over the face of the ocean, and thus rushing forward to imminent danger or probable destruction. At sunset, accordingly, the sky became cloudy and threatening, and in the course of the night it blew a very hard gale, which afterwards continued for three days successively. This gale completely dispersed the fleet of boats, and it was not without the utmost difficulty that many of them reached the various creeks of the island. It is believed no lives were lost on this occasion, but the boats were damaged, much tackle was destroyed, and the men were unnecessarily exposed to danger and fatigue. During the same storm, it may be remarked, thirteen vessels were either totally lost or stranded between the Isle of Anglesea and St. Bee's Head in Lancashire. Mr. Stevenson remarks, how much it is to be regretted that the barometer is so little in use in the mercantile marine of Great Britain, compared with the trading vessels of Holland, and observes, that although the common marine barometer is perhaps too cumbersome for the ordinary run of fishing and coasting vessels, yet Adie's sympiesometer is so extremely portable that it might be carried even in a Manx boat. Each lot of such vessels has a commodore, under whose orders the fleet sails ; it would therefore be a most desirable thing that a sympiesometer should be attached to each commodore's boat, from which a preconcerted signal of any expected gale or change of weather as indicated by the sympiesometer could easily be given."

THE HABITS OF FISHES.

The following notes as to the habits of fish may prove of interest to the naturalist :—

"It has often been observed in the course of the Bell Rock operations, that during the cold weather of spring and autumn, and even at all seasons, in stormy weather, when the sea is much agitated by wind, the fishes disappear entirely from the vicinity of the rock, probably retreating into much deeper water, from which they do not seem to return until a change of weather has taken place; so much was this attended to by the seamen employed on this service, that they frequently prognosticated and judged of the weather from this habit of the fishes as well as from the appearance of the sky."

"It was a general remark at the Bell Rock that fish were never plenty in its neighbourhood, excepting in good weather. Indeed, the seamen used to speculate about the state of the weather from their success in fishing. When the fish disappeared at the rock, it was considered a sure indication that a gale was not far off, as the fish seemed to seek shelter in deeper water, from the roughness of the sea, during these changes of the

2 B

weather. This evening, the landing master's crew brought
to the rock a quantity of newly caught cod fish, measuring from fifteen to twenty four inches in length. The
membrane called the *sound*, which is attached to the
backbone of fishes, being understood to contain, at different
times, greater portions of azote and of oxygen than common air, the present favourable opportunity was embraced
for collecting a quantity of this gas in a drinking glass
inverted into a pail of salt water. The fish being held
under this glass as a receiver, their bladders were punc-
tured, and a considerable quantity of gas was thus collected.
A lighted match was afterwards carefully introduced into
the glass, when the gas exhibited in a considerable degree
the bright and luminous flame which an excess of oxygen
is known to produce."

On showing this extract to my friend Dr. P. D.
Handyside, who has contributed some interesting papers to
the Royal Society of Edinburgh on the Polyodon gladius,
he writes :—"Biot and De La Roche found that the proportion of oxygen in the air bladder increases with the
depth of the water in which the fish usually lives, from
a small quantity up to 87 per cent. Biot found in the
deep Mediterranean fishes 87 parts of oxygen, nitrogen,
and carbonic acid. Humboldt found in the electrical eel
96 parts of nitrogen and 4 only of oxygen. No hydrogen
has ever been detected in this organ. In the air bladder
of marine fishes oxygen predominates, and in that of
fresh-water fishes nitrogen. No air sacs exist in rays,
flounders, sole, turbot, and others which lie at the bottom."

Dr. Handyside adds : " The extract shows with what a

practical and accurate mind your father was endowed, and I think, in justice to him, you should give his observations."

I also communicated Mr. Stevenson's papers on fishings to the Honble. B. F. Primrose, C.B. (Secretary to the Fishery Board : Scotland), who has kindly sent me a letter explaining why the progress of the fishings in the Shetland Islands is slow, from which I give a few extracts :—

"I have read with great interest your father's notes upon the fisheries of Scotland. They bear distinctly the impress of that practical and accurate mind with which he is described as having been endowed. It is also pleasant to see that his mind went a great deal further, and grasped the application of science to solve the mystery of fishings.

"He seems to have overlooked, as was universal in his day, that the secret of fisheries is not the presence of fish but the certainty of markets. Samuel Laing of Orkney, to whom he refers, was, I think, the first that struck this key note of truth. The Dutch came here and fished for herrings because they could not fill their vessels fast enough for the markets behind them in Holland. The Shetlanders did not fish for herrings because they had no remunerative market for them, but they fished, and fished boldly, where they had one, viz., for the whales of the Arctic Regions. They might have brought the herring home from off their own coasts and got nothing for them, but they could not bring the whale oil home without a secured profit.

"The same thing obtains still. Shetland, from its position, cannot compete with the mainland of Scotland either in the home market or in the great continental markets for herrings; but it yields large supplies of cod, ling, and tusk, for which it pushes distant adventures to Iceland and the Faroe Isles."

CHAPTER XV.

MARINE SURVEYING.

MODERN engineers who have practised only under the benign reign of Ordnance Surveys and Admiralty Charts, can have no idea of the toil their predecessors underwent in procuring data for their designs and reports; and I am safe in saying that Mr. Stevenson was of all others the engineer to whom in his sea coast practice, such useful aids would have been of the very highest value.

For example, before he could tell, with the exactness he desired, the distance between the Bell Rock Lighthouse and the shore, he had, in absence of any reliable information, to undertake a pretty extensive trigonometrical survey of the coast, involving the measurement of a *base line* upwards of two miles in length—a most "laborious operation," he observes, in which his assistants were aided by six sailors from the lighthouse tender.

Again, to show the difficulty in determining the best site for a lighthouse in those early days, before an accurate Government survey of the coast line had been made, I give from Mr. Stevenson's Journal the following notes of his observations to determine the best site for a lighthouse at Kinnairdhead in Aberdeenshire. I give them *at length*, as jotted down at the time, for they may perhaps lead young engineers of the present day to be thankful that, in most cases at least, they are not, from want of accurate coast surveys and soundings, left to resort altogether to their own resources in getting the information they require. But I think they are specially worthy of record as showing the extreme care bestowed by Mr. Stevenson in getting the data to enable him to determine the exact positions of the several lighthouses *he designed.* His Journal says :—

"*First.*—I caused a mast to be erected upon the top of Kinnaird-head Castle or Lighthouse, making its extreme height from the ground 100 feet.

"Got the yacht under weigh, and having a careful pilot on board, I sailed for Rattray Head, and there observed the mast over the land of Cairnbulg, it being then high water, or twenty minutes past 7 P.M. With the parapet of the lighthouse in view, have eight fathoms water off the head, which bore W.N.W. Run in upon the head with flag upon the mast seen over the land till seven fathoms water, when the flag disappeared. Then leave the vessel and sound from the boat, and have 6 fathoms, 5, 5, $4\frac{1}{4}$, 3, 2, 1 fathom, and lastly $3\frac{1}{2}$ feet. Return to the ship in a more southerly direction, and have 3 feet, 1 fathom, 2, $2\frac{1}{4}$, $2\frac{3}{4}$, $3\frac{1}{2}$, 4, $4\frac{3}{4}$, 5, $5\frac{1}{2}$, $6\frac{1}{2}$, and 7 fathoms. All these soundings rocky bottom.

"With the Windmill near Peterhead on with Stirling hill, and Monument hill on with the rounded Sandy Down of Rattray, and the parapet of Kinnairdhead Lighthouse seen over Cairnbulg land, you are in 8 fathoms water off Rattray Briggs, which lie about $\frac{1}{4}$ of a mile to the southward of the Sandy Down.

"Wait off the Briggs till the light was seen, then stood in upon the Briggs till the light was shut in by the land of Cairnbulg, and at that moment had 8 fathoms water, so that at present the light forms an excellent direction for Rattray Briggs.

"Find that the lightroom is seen fully from the yacht's deck in 8 fathoms water off Rattray Briggs, that the flag upon the mast-head is seen in 6 fathoms water—high water spring tides. Ship then bearing from the head E.S.E. and W.N.W., distant about one mile from the shore, where a man is distinctly observed at a boat in the twilight.

"*Secondly.*—Remove the mast from the castle or lighthouse on the morning of the 15th to Cairnbulg, and elevate a flag to the height of 86 feet from the ground, or 97 feet from high water mark, at the distance of about 100 yards from the high water mark at the point connected with Cairnbulg Briggs.

"The yacht lying off or to the westward of the Briggs, was got under weigh at 2 A.M. of the 16th, and beat up the north shore as far as Rosehearty, and there observed the flag over the land. Found off Rosehearty that the flag was just hid by the highest inequalities of the land to the southward of the Castle, and that it appeared at the lower or flat places sometimes in sight 20 feet above the land, and at other places intercepted by the land and houses of the town, amongst which it often appeared and disappeared. The range of the flag along the land was as far as Mr. Dalrymple's house when it was time to put about, having there three fathoms at nearly low water.

"After completing the observations in this direction, sailed along the shore southwards to Rattray Briggs. Find that Inverallochy head, south-eastward of the town of Cairnbulg, is the eastmost point on this coast, but, being at a distance from the foul ground of Cairnbulg, would make a less desirable point than Cairnbulg.

"Off Rattray, in eight fathoms water, begin to lose sight of the lantern on Kinnairdhead Castle as before. See the mast and flag at Cairnbulg a considerable way up the country over the lands of Inverallochy. See the flag, standing in upon Rattray to five fathoms water at half tide, lose it, and then stand for Fraserburgh.

"As the result of these trials, find that Inverallochy head or point is the most eastern or projecting point of land upon that coast, that Cairnbulg is the next projecting point. The former lies between the points of danger, viz., Rattray and Cairnbulg.

"Find that if the light were to be moved to a more southern situation, it would be better on either of the above places than Rattray Head, which would entirely remove its usefulness from the Moray Firth.

"Find that in the event of two lights for this coast, the one ought to be at Kinnairdhead, and the other upon the Cock Inch at Peterhead.

"Under all the circumstances of the case, find that it would be most advisable to erect a new lighthouse at Kinnairdhead, about 100 yards more to the eastward than the Castle stands, and erect it about twenty or thirty feet higher than the Castle. This, with a better light, would perhaps answer the general purposes of the coast better than a single light placed on any of the other stations along this coast."

After perusing this extract, the reader, I think, will not be surprised to find Mr. Stevenson making an urgent appeal on behalf of all interested—Seamen—Fishermen, and Engineers, for a Government Survey and "Sailing Directions" of the intricate navigation of the shores of Scotland, which he did in the following terms :—

"The attention which Government has long paid to the improvement of the Highlands and Islands of Scotland, in connection with the British Fisheries, has been attended with the best effects in the country at large. It is much to be wished that these shores were rendered more accessible to the mariner.

"The marine survey of the Highlands by Murdoch Mackenzie, undertaken by order of the Lords of the Admiralty, may be considered as the first grand step towards the improvement of the Highlands, and next to that the later institution of the Northern Lighthouses. By means of these the fisher may find his way from loch to loch, and the mariner bound over seas, instead as formerly of holding a course without the Lewis Islands, can now find his way through the Sounds, and in adverse winds take shelter in safe harbours, instead of being exposed to the boisterous seas of the Atlantic Ocean ; these charts and lighthouses have in many points of view contributed to the improvement of the Highlands, and to the present flourishing trade carried on through these Sounds from Liverpool, etc., to the northern continent of Europe. However, from the extensive range of coast which these charts include, together with the prodigious number of extensive lochs and small islands, it was impossible that any first survey could be made so accurate as to supply the place of pilots, where there are neither landmarks to characterise the coast, nor beacons or buoys to point out the situation of sunk rocks ; and although these charts have certainly contributed much to the facility and security of the navigation of the Highlands, yet no one will say that they are free from imperfections, and their incommodious size and high price are insurmountable bars to their general utility, thereby rendering them impracticable for the use of small vessels, so that they are only to be found in the cabins of large vessels, where large accommodation affords room to unfold them, but even here also the price forms an objection, as the charts are always *found* by the shipmaster.

"Nothing therefore can be more necessary or essential to the improvement of the Highlands and Islands of Scotland, than an accurate survey of the fishing grounds, lochs, and harbours, upon a scale considerably larger than Mackenzie's charts, given in the form of a book of the size of a large quarto, containing only the lochs, etc., interleaved with printed directions and descriptions of each chart or harbour, which book of charts, accompanied with a general chart would sufficiently guide the mariner and fisher in their several pursuits.

" With regard to an accurate survey of the lochs and harbours in the Highlands published in the most commodious form for the use of small vessels, such an undertaking would require to be sanctioned in a manner similar to the survey undertaken by Murdoch Mackenzie, and though in process of time the sale of these charts might produce a considerable return to those concerned with it, yet the time and attention which such (with a laborious number of soundings) must occupy would certainly require that those concerned in the undertaking should be put in possession of certain sums of money to enable them to go on with that deliberation which is essential to accuracy, and this encouragement should be the more considerable that the charts might be procured to the public at a moderate price."

This Memorial, written in 1803, was intended for and in some shape communicated to the Admiralty, and was followed by good results.

In " A Memoir of the Hydrographical Department of the Admiralty," published in 1868,[1] are the following remarks :—" It was about this time," 1810, " that the Admiralty first conferred on the Hydrographer the privilege of selecting a surveyor for the *home coasts.* Singular as it may appear, the Hydrographer had at this time great difficulty in finding a naval officer competent to fill the position, or who was acquainted with anything beyond surveying by common compass. At length, however, about 1811, Mr. George Thomas, a master, was selected " for home service. The Memoir

[1] A Sketch of the Institution and the progress of the Hydrographical Department of the Admiralty, from its first establishment in the year 1795.

2 C

also states that at the same time the Hydrographer appointed to foreign service Mr. Beaufort, afterwards Sir Francis Beaufort, the eminent Hydrographer to the Admiralty, who was, all his life, Mr. Stevenson's intimate friend and constant correspondent.

There is therefore, I believe, no reason to doubt that Mr. Stevenson's original appeal and subsequent personal friendly and free intercourse with the officials of the Admiralty led to the establishment, on a *systematic footing*, of our Government "Admiralty Survey," which, as all engineers know, indicates with marvellous accuracy and detail every shoal, sunken rock, and sounding on the coasts of Great Britain and Ireland; and from which the "Admiralty Sailing Directions" have been prepared with such discernment and care that the whole system of our coast survey may now be said to have attained perfection.

With Colonel Colby, also, of the Royal Engineers, who was Director of the Ordnance Survey, Mr. Stevenson regularly corresponded, being no less interested in the progress of the great national work so successfully carried on under his charge.

CHAPTER XVI.

CONTRIBUTIONS ON ENGINEERING AND SCIENTIFIC SUBJECTS.

Contributions to *Encyclopædia Britannica* and *Edinburgh Encyclopædia*—The alveus or bed of the German Ocean—Sectio planography—Wasting effects of the sea at the Mersey and Dee—Density of fresh and salt water—The Hydrophore.

WE have seen that Mr. Stevenson's college education was mainly, if not altogether, due to his own thirst for knowledge, and his education being voluntarily undertaken, could hardly fail to issue in good results. That his early studies were of incalculable value to him no one can doubt; and his own conviction of this may explain the solicitude with which, in after life, he impressed on his sons the extreme importance of being properly grounded in every branch of study, *scientific* and *practical*, which a well trained engineer has to call to his aid in the practice of his profession.

Fortified by this valuable training, Mr. Stevenson had also that unselfish love of his profession which alone can move a man to give the results of his experience freely to others, and this he did to the *Edinburgh Encyclopædia* and the *Encyclopædia Britannica*, in articles on "Roads," "Lighthouses," "Railways," "Dredging," "Blasting," and other engineering subjects.

But he did not confine his literary labours to matters purely professional. His love for nature in all its aspects led him also to make communications to the Scientific Journals of the day on subjects of more general interest. Of these his papers " On the Alveus or Bed of the German Ocean," in which by an investigation of many evidences he is led to the conclusion that the sea is gradually encroaching on the land, may be quoted as an example.

Mr. Stevenson's first communication on this subject was published in 1816, in vol. ii. of the Wernerian Transactions, in which he gives examples, from actual observation, of the wasting effects of the sea on various parts of the coasts of the British Isles. His second communication was made to the Wernerian Society in March 1820, and published in the *Edinburgh Philosophical Journal* of that year.

In the fifth edition of Baron Cuvier's " Essay on the Theory of the Earth," reference is made to Mr. Stevenson's theory. His papers are several times quoted in Lyell's *Principles of Geology*, and the General Committee of the British Association at York in 1834 passed a resolution, " that Mr. Stevenson be requested to report to the next meeting upon the waste and extension of the land on the east coast of Britain, and upon the general question of the permanence of the level of the sea and land, and that individuals who may be able to supply information upon the subject be requested to correspond with him."

Without discussing in how far Mr. Stevenson's theory may be sound (for on such questions it is notorious that the views of geologists do not always coincide), it cannot be denied that his mode of dealing with the subject is original and interesting, and as the papers are not now accessible to the general reader, it may be excusable to give one of them *in extenso*. I also notice another feature which gives interest to the subject. In his illustrations he adopted a mode of representation which was peculiarly suitable for the object in view. It will be seen from Plate XII. that the sections are laid down on what is now known by engineers as *sectio planography*, which it is believed was used for the first time in illustrating this paper.

" On the Bed of the German Ocean, or North Sea. (Read before the Wernerian Natural History Society, 25th March 1820.)"

"The efforts of man in exploring the more occult processes of nature are necessarily much circumscribed, especially when his attempts are directed to the investigation of regions which his senses cannot penetrate. It has accordingly been with the utmost difficulty that his exertions have been rendered in any degree successful in prying into the bowels of the earth, or in his endeavours to ascend to the aërial regions. In proof of this, the limited excavations even of the most extensive mining works, have required the lapse of ages, and the powerful stimulus of commercial enterprise, for their accomplishment. From these the philosopher has not hitherto derived much light, to enable him to compare the theories which have been assigned by geologists to account for the various and discordant appearances of the structure of the globe.   It has also been with much difficulty, and at no small personal hazard, that the philosophical inquirer has ventured to climb the highest mountains, to examine into the phenomena of the atmosphere.   The balloon has indeed enabled us to attain still higher points of elevation; but as yet we do not seem to have made proportional progress in knowledge.   In all such attempts to ascend the greatest heights or penetrate the deepest excavations, we still breathe in our own element, though under different modifications.   If, however, we would explore the depths of the Ocean, we immediately encounter an element to which the organisation of our lungs is not at all adapted; the density of air, compared with water on a level with the surface of the sea, being in the ratio of one to about 850; and our difficulties must consequently increase in a very rapid proportion.   Here therefore we are unavoidably left to conjecture on many points of our inquiries regarding this highly interesting subject.   Even the ingenious contrivance of the diving bell contributes but little towards our investigations for ascertaining the nature of the bottom of the sea, at least to any considerable depth, on account of the difficulty of its application in situations exposed to stormy weather, and also of the increasing ratio of the pressure of the fluid as we descend.   This curious machine, it is believed, was invented and employed, about the year 1720, by a Captain Rowe for raising the wreck of ships upon the coast of Scotland; and in the year 1778, the active mind of Smeaton first applied it to the operations of the engineer.

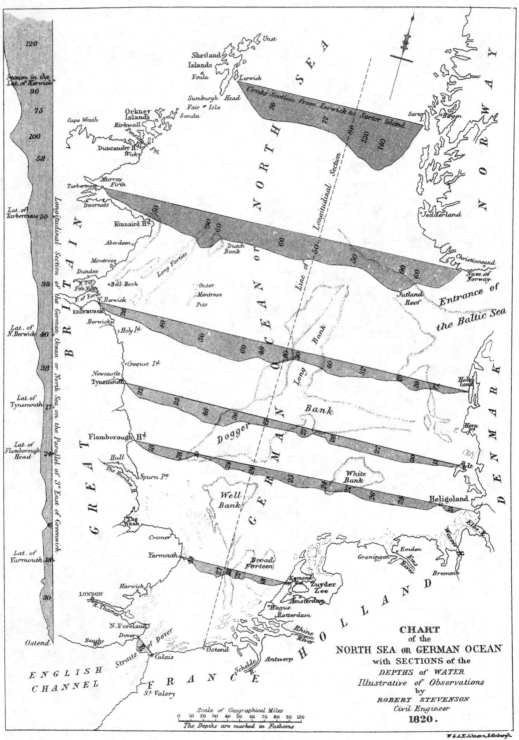

CHART
of the
NORTH SEA OR GERMAN OCEAN
with SECTIONS of the
DEPTHS of WATER
Illustrative of Observations
by
ROBERT STEVENSON
Civil Engineer
1820.

Scale of Geographical Miles
0 10 20 30 40 50 60 70 80 90 100
The Depths are marked in Fathoms

W & A.K. Johnston, Edinburgh.

" Our knowledge of the bottom of the ocean, therefore, remains still very imperfect, and, with little exception, the simple apparatus of the mariner, consisting of a plummet and line, continues to be chiefly in use for ascertaining the depth of the sea and the nature of the ground. With these, and the addition of a little grease applied to the lower extremity of the plummet, which strikes against the bottom, we learn the quality of the soil, though imperfectly, by the particles which adhere to the grease. What the navigator has yet been able to discover regarding the depth and the nature of the bottom of the German Ocean, I shall now endeavour to notice, being myself enabled to offer the result of a pretty extensive acquaintance with this field of inquiry.

" It may be necessary to promise, in treating of a subject so extensive, and in comparing great things with small, that we are obliged to speak of the North Sea as a bay or basin, and of the immense collection of débris which we meet with, extending over a great proportion of its bottom, under the common appellation of sand banks. We must also be allowed to consider the undulating line, or the irregularities of the bottom, to arise chiefly from the accumulation of deposited matters ; and in most of the situations connected with these banks, we are supported and borne out in this conclusion, by their local positions relatively to the openings of firths, and the line of their direction in regard to the set or current of the ebb tide.

" The accompanying map (Plate XII.) of the eastern coast of Great Britain, with the opposite Continent, though upon a small scale, exhibits numerous soundings of the depth of the German Ocean; and the sections delineated on it will perhaps be found to give a pretty distinct view of the subject. This chart extends from the coast of France, in latitude 50° 57′ to 61° N. On the east, this great basin is bounded by Denmark and Norway, on the west by the British Isles, on the south by Germany, Holland, and France, and on the north by the Shetland Islands and the Great Northern or Arctic Ocean. The term *German Ocean*, though in very common use, is certainly not so comprehensive in its application to this great basin as that of *North Sea*, now more generally used by the navigator. The extent of this sea from south to north, between the parallels of latitude quoted above, is 233 leagues, and its greatest breadth from west to east, reckoning from St. Abb's Head, on the coast of Scotland, to Ringhjoöbing Fiord, on the opposite shore of Denmark, is 135 leagues. The greatest depth of the water in this basin seems to be upon the Norwegian side, where the soundings give 190 fathoms; but the mean depth of the whole may be stated at only about 31 fathoms.

"To be more particular with regard to the depth of the German Ocean, or North Sea, it will be observed by the sections and soundings marked upon the chart, that the water gradually deepens as we sail from south to north. The first of these sections which we shall notice is on the parallel of three degrees of east longitude, running from Ostend to the latitude of the northmost of the Shetland Islands, being an extent of 227 leagues. The depth, as will be seen from this section (which, to avoid confusion in the body of the chart, is traced along the western side of it), varies rather after an irregular progression, from 120 fathoms towards the northern extremity of this sectional line, to 58, 38, 24, and 18 fathoms, as we proceed southwards, to within five miles of the shore, nearer which we do not approach in our remarks regarding the soundings. Notwithstanding the irregularity of the depth from the occurrence of numerous sandbanks, it is curious to observe the increase upon the whole as we proceed from south to north, by which this sea exhibits all the characteristic features of a great bay, encumbered with numerous sandbanks.

" In the same manner, though not strictly connected with our present purpose, we may observe that the English Channel deepens progressively from Dover to its entrance, formed by the Land's End of England and the Isle of Ushant, on the coast of France ; so that the Strait between Dover and Calais may be said to form a point of partition between two great inclined planes, forming the bottom of these seas.

" Besides the longitudinal, or north and south sectional line described above, we have also six other sections delineated in an easterly and westerly direction, across the accompanying chart, which are as follow. One between the Shetland Islands and the coast of Norway ; a second between Tarbetness in Ross-shire and the Naze of Norway ; a third extends from the Firth of Forth to the coast of Denmark ; a fourth from the mouth of the river Tyne to Sylt Island, also in Denmark ; a fifth from Flamborough Head, in Yorkshire, to the mouth of the

River Elbe ; and the sixth is from Yarmouth to Egmond-op-Zee, on the coast of Holland. Other sections of this sea have also been made, which include the general elevation of the land, as, for example, one of these extends from Holland across the German Ocean to the Thames, and through the interior of the country to the Bristol Channel ; then crossing St. George's Channel, this sectional line passes through the southern extremity of Ireland, and falls into the Atlantic Ocean ; but this will be more particularly noticed, when I come to speak of the bed of the English Channel, in a future paper.

"On examining the accompanying cross sections of the depths of water on the same parallel they will be found to vary considerably. It may, however, be stated as a general conclusion, that there is a greater depth of water on the eastern and western sides of the German Ocean than in its central parts, and that, upon the whole, it is deeper on the British than on the continental shores, the coast of Norway excepted.

"We have already observed, that this sea is much encumbered with sandbanks, or great accumulations of débris, especially in the middle or central parts, and also along the shores towards what may be termed the apex of the bay, extending from the river Thames along the shores of Holland, etc., to the Baltic. One of these great central banks, delineated on the chart, and known to mariners as the Long Forties, trends north-east in the direction of the ebb tide from the entrance of the Firth of Forth no less than 110 miles, while the Denmark and Jutland banks may also be traced on the chart from the entrance of the Baltic, upwards of 105 miles in a north-western direction. Besides these, we have also another great central range of banks, which is crossed by no fewer than four of our sectional lines. These are known under the common appellation of the Dogger Bank, which is subdivided by the navigator into the Long Bank, the White Bank, and the Well Bank, including an extent of upwards of 354 miles from north to south. There are also a vast number of shoals and sandbanks, lying wholly to the southward of our section, between Flamborough Head and Heligoland. Altogether, therefore, the superficies of these extensive banks is found to occupy no inconsiderable portion of the whole area of the German Ocean; the surface of which, in making these investigations, has been estimated to contain about 153,709 square miles, while the aggregate superficial contents of the sandbanks alone amount to no less than 27,443 square miles, or include an area of about $5\frac{3}{4}$ of the whole surface of the North Sea.

"But to render these dimensions a little more familiar by comparison, we may notice, that the Island of Great Britain contains about 77,244 square miles, being not quite one half of the area of the North Sea; so that the area of the sandbanks bears a proportion equal to about one third of the whole *terra firma* of England and Scotland; and they are, therefore, perhaps, far more considerable in their extent than has been generally imagined.

"In speaking of the dimensions of sandbanks situate in the middle of the ocean, we are aware that great allowance must be made in forming a proper estimate of their extent, especially in speaking of their cubical contents. From a vast number of observations and comparisons relative to this subject, I have, however, been enabled to determine, that the average height of these banks measures about seventy-eight feet, from a mean taken of the whole. In ascertaining their height above the surrounding bottom, the measurement has been taken from the general depth around each respectively. Now, upon taking the aggregate cubical contents of the whole of these immense collections of débris, supposing the mass to be uniformly the same throughout, it is found to amount to no less a quantity than 2,241,248,563,110 of cubic yards, being equal to about fourteen feet of the depth of the whole German Ocean, or to a portion of the firm ground of Great Britain, on a level with the sea, taken twenty-eight feet in perpendicular height or depth, supposing the surface to be a level plane.

"These calculations at least tend to show that an immense body of water must be displaced, in consequence of these banks occupying so very considerable a proportion of the bed of the North Sea, the unavoidable effect of which must give a direct tendency to the tidal waters, and the flux produced by storms in the Atlantic, to overflow the bed of the German Ocean, in the same manner as if stones or other matter were thrown into a vessel already nearly brimful of water. This may further be illustrated by considering the actual state of any of the great inland lakes, as those of Geneva, Lochness, Lochlomond, etc., which for ages past have been receiving the débris of the surrounding mountains. We must doubtless allow that they contain a smaller portion of water, or are actually of a less depth than they were at an earlier period of the history of the globe. Accordingly, from inquiries, which, in the prosecution of this subject, I have been led to make regarding the two last mentioned lakes, it has satisfactorily appeared that their waters are subject to overflow or rise upon their banks. On Lochlomond, in particular, the site of a house at the village of Luss was pointed out to me, which is now permanently under *the summer water mark,* while the gable of another house in its neighbourhood is in danger of being washed down by the increase of the waters of the loch. Whether this striking appearance is to be attributed wholly to natural causes, or partly to artificial operations upon the bed of the river Leven, flowing from the loch, I have had no opportunity of inquiring. But the great bench or flat space round the margin of the loch, which is left partly dry during summer, forms altogether such a receptacle for débris as to be sufficient to affect the surface of the loch, and indeed permanently to raise its waters. We also infer, though by a different process, that the constant deposition going forward in the bed of the German Ocean must likewise displace its waters, and give them a tendency to enlarge their bed and to overflow their banks or boundary.

" In this view of the subject, it will appear that we have not only to account for the supply of an immense quantity of débris, but we must also dispose of the water displaced by the process of deposition which is continually going forward at the bottom of the ocean.

" With regard, then, to the supply of the débris of which
these banks are composed.—We find that a very great
portion of it consists of siliceous matters in the form of
sand, varying in size from the finest grains to coarse
bulky particles, mixed with coral and pounded shells, the
quantity of these calcareous matters being altogether
astonishingly great; and being specifically lighter than
the particles of sand, the shells generally cover the surface
of these sunken banks.   With regard to the vast collection
of siliceous particles connected with the banks, our surprise ceases when we consider the receptacle which the
North Sea forms, to an almost unlimited extent of drainage
from the surrounding countries, on which the change of
the seasons, and the succession of rain and of drought
upon the surface of the earth, are unceasingly producing
their destructive effects.   All have remarked the quantity
of mud and débris with which every rill and river is
charged, even after the gentlest shower; especially
wherever the hand of the agriculturist is to be found.
His labours in keeping up the fertilising quality of the
ground consist in a great measure in preparing a fresh
matrix for the chemical process or the germination of the
seeds of the earth, in lieu of that portion of the finely
pulverised soil which the rains are perpetually carrying to
the sea, as the grand receptacle and storehouse of nature
for these exuviæ of the globe.   From the effect of rills
and rivulets, we should, perhaps, rather be apt to expect
a greater deposition in the bed of sheltered bays and arms
of the sea than we really observe.   So that we can readily
believe that the quantity of débris, even for a single year,

along such an extent of coast, may bear some consideration in respect to the bed of the German Ocean ; what, then, must these effects produce in the lapse of ages ?

" Whatever be the cause, the fact is certain, that on almost every part of the shores of Great Britain and Ireland, and their connecting islands, from the northernmost of the Shetland to the southernmost of the Scilly Islands, and also upon the shores of Holland, and part of France, particularly in the neighbourhood of Cherbourg, this wasting effect is going forward. These shores I have myself examined. But my inquiries have not been confined to the coasts which I have personally visited, having also, through the kind attentions of some nautical friends, been enabled to extend my investigations even to the remotest parts of the globe. The general result has been, that equally in the most sheltered seas, such as the Baltic and Mediterranean, and on the most exposed points and promontories of the coasts of North and South America, and the West India Islands, abundant proofs occur, all tending to show the general waste of the land by the encroachments of the sea. Such wasting effects are quite familiar to those locally acquainted with particular portions of the shores ; and I have often received their testimony to these facts, as the sad experience of the removal of buildings, and the inundation of extensive tracts of land by the encroachment of the sea.

" Indeed, by a closer inquiry into this department of the subject, we shall, perhaps, find ourselves rather at a loss to account for the *smallness* of the quantity of this deposition, considering the waste which is constantly

going forward in the process of nature, and even be led
to seek for its wider distribution over the whole expanse
of the bed of the ocean, as has been supposed in that
theory of the globe, so beautifully and so ably defended
by our late illustrious countryman Professor Playfair.

"One of the most striking and general examples of this
kind may perhaps be found in the abrupt and precipitous
headlands and shores which we everywhere observe along
the coast, and which we suppose to have once been of the
same sloping form and declining aspect with the contiguous land. In the production of these effects alone, an
immense quantity of débris must have been thrown into
the bed of the ocean. The channels which are cut by
the sea in the separation of parts of the mainland, and
the formation of islands, no doubt make way for a considerable portion of the displaced fluid; but still these
channels, when filled with water, come far short, in point
of bulk, when compared with the portions of the elevated
land which are thus removed. Now, it has been alleged
by some, that while the land is wasting at certain points,
it is also gaining in others; and this is a state of things
which is freely admitted to take place in various quarters;
yet these apparent acquisitions are no more to be compared with the waste alluded to, than the drop is to the
water of the bucket. But accurate observations regarding
the formation of extensive sandbanks, and the accumulation of the débris, of which they are formed, are not to be
made in a few years, perhaps not in a century, nor indeed
in several centuries; for although the short period of the
life of man is sufficient to afford the most incontrovertible

proofs of the waste of the land where we become observers, yet when we extend our views to the depths of the ocean, and speak of the events and changes which are there going forward, we must not be supposed to set limits to time.

" We have many convincing proofs in the natural history of the globe, that the sea has at one time occupied a much higher elevation than at present. On the banks of the Firth of Forth, near Borrowstounness, for example, I have seen a bed of marine shells, which is several feet in thickness, and has been found to extend about three miles in length, and which is now situate many feet above the present level of the waters of the Forth. A recent illustration of this subject occurred also in the remarkable discovery of the skeleton of a large whale, found in the lands of Airthrey, near Stirling,—the present surface of the ground where the remains of this huge animal were deposited, having been ascertained (by my assistants, when lately in that neighbourhood) to be no less than twenty-four feet nine inches above the present level of the Firth of Forth at high water of spring tides. Now, whether we are to consider these as proofs of the higher elevation of the waters of the ocean in the most general acceptation of the word, at a former period, I will not here attempt to inquire. But aside from these anomalous appearances, there is reason for thinking that the waters of the higher parts of the Firth of Forth, like those of the Moray Firth, may, at one time, have formed a succession of lakes, with distinct barriers, as we find in the case of Lochness, and the other lakes forming the track of the Caledonian Canal. My object on the present occasion,

2 E

however, is simply to notice the wasting effects of the
North Sea upon the surrounding land, its deposition in
the bottom of the sea, and the consequent production
of surplus waters at the surface, and to endeavour to
account for these appearances consistently with the laws
of nature.   The opinion accordingly which I have formed,
and the theory which I have humbly to suggest (for I
am not aware that this subject has been before particularly
noticed) is, that the silting up of the great basin of the
North Sea has a direct tendency to cause its waters to
overflow their banks.

  " Referring to the chart, we find that the North Sea is
surrounded with land, excepting at two inlets or apertures, the one extending about 100 leagues, between the
Orkney Islands and the Norwegian coast, and the other
between Dover and Calais, which is of the width of seven
leagues.   The aggregate *waterway* of these two passages
forms the track for the tidal waters, and also for the
surplus waters produced during storms which affect the
Atlantic and Arctic Oceans.   It is also obvious that this
waterway must remain nearly the same, and admit a
constant quantity; or, to speak more correctly, by allowing these inlets to follow the general law, they must be
enlarged by the waste or wearing of their sides, in a
ratio perhaps greater than the silting up of the bottom in
those particular parts, while the interior and central
portions of the German Ocean are continually acquiring
additional quantities of débris, along with the drainage
water of the widely surrounding countries.   If therefore
the same, or a greater quantity of tidal and surplus

waters continue to be admitted from the Atlantic and Arctic Seas into this great basin, where the process of deposition is constantly going forward, it is evident that the surface of the German Ocean must be elevated in a temporary and proportionate degree, and hence the production of those wasting and destructive effects which are everywhere observable upon its shores.

" This reasoning is also applicable, in a greater or less degree, to all parts of the world ; for as the same cause everywhere exists, the same effects, when narrowly examined, must everywhere be produced. In the Southern or Pacific Ocean we have wonderful examples of great masses of land formed by madrepores and extensive coral banks, which in time assume all the characteristic features of islands. These occupy considerable portions of the watery bed of the ocean, and displace corresponding portions of the fluid. Immense quantities of mud are also said to be deposited in the Yellow Sea of China, in the great deltas formed at the mouths of the Ganges, the Plate, the Amazon, the Mississippi, the St. Lawrence, the Nile, the Rhine, and other large rivers, whose joint operations, both at the surface and bottom of the ocean, are continually carrying forward the same great process of displacing the waters of the ocean ; for it matters not to this question whether the débris of the higher country which is carried down by the rains and rivers, or is occasioned by the direct waste produced by the ocean itself on the margin of the land, be deposited at the bottom or surface of the ocean, it must still be allowed to displace an equal or greater bulk of the fluid, and has therefore a direct tendency to produce the derangement which we are here endeavouring to describe.

"A striking illustration of this doctrine may be drawn from M. Girard's able and ingenious observations on the delta of Egypt, made in 1799, and published in the *Mem. de l'Acad.* for 1817, in a memoir *Observations sur la Vallée d'Égypte, et sur l'exhaussement séculaire du sol qui la recouvre.* It appears that the whole soil of the "Valley of the Nile" is very considerably increased by the alluvium deposited annually by the inundations of the Nile, as ascertained by the marks on some ancient Nilometers and statues, the dates of which have been traced and compared by Girard, with the corresponding historical periods. In the quarter of Thebes, where the statue of Memnon is erected, the increase of the soil since the commencement of the Christian era is $1^m\cdot924$ (6 feet 3·7 inches), or this process may be stated as going forward at the rate of $0^m\cdot106$ (4·17 inches) in the course of each century. The magnitude of the deposits at the mouths of the Nile, in the bed of the Mediterranean, appears to be no less surprising. It is remarked that the Isle of Pharos, which in the time of Homer was a day's journey from the coast of Egypt, is now united to the continent.

"If, then, we compare these effects with the same process, going forward in a certain proportionate rate over all parts of the globe, and where the same facilities for these depositions being made on firm ground are not afforded, we shall find that the quantity of deposit in the bottom of the ocean must be so considerable as to affect the level of the waters of the ocean.

"In thus disposing of the waste of the surrounding land beyond the accumulation of the sunken banks in the German Ocean, we are not left at any loss for a distributing cause, as this is provided by the tides and currents of the sea; and with regard to their action we have many proofs, even at very considerable depths, by the breaking up of the wrecks of ships, the occasional drift of seaweed, and also drift timber, nuts, etc., into regions far distant from those in which they are spontaneously produced. The dispersion of fishes, evinced by their disappearance from the fishing grounds in stormy weather, tends to show the disturbance of the waters of the ocean to the depth of thirty or forty fathoms. This observation I have frequently had an opportunity of making near the entrance of the Firth of Forth. Numerous proofs of the sea being disturbed to a considerable depth have also occurred since the erection of the Bell Rock Lighthouse, situate upon a sunken rock in the sea, twelve miles off Arbroath, in Forfarshire. Some *drift stones* of large dimensions, measuring upwards of thirty cubic feet, or more than two tons weight, have, during storms, been often thrown upon the rock from the deep water. These large boulder stones are so familiar to the lightkeepers at this station as to be by them termed *travellers*. It is therefore extremely probable, that a large portion of the débris is carried down with the drainage water of the higher country, as before noticed, and ultimately washed out of the North Sea into the expanse of the ocean.

"The question which naturally arises as to the result

of all this waste or transposition of the solid matters of a
large portion of the globe, is to inquire what has become
of the body of water displaced by this wasting process.
Without attempting to go into all the minutiæ of this
part of the subject, I shall here briefly observe, that
there seems to exist (if I may be allowed so to express
myself) a kind of compensating arrangement between
the solid or earthy particles of the globe in the one
case, and the waters of the ocean in the other.   Thus
by the process of evaporation, and the universal application of water, which enters so largely, in its simple or
chemical state, into the whole animate and inanimate
creation, the surface of the ocean may be kept nearly at
a uniform level.   Phenomena of this description are, no
doubt, difficult in their solution upon the great scale,
being met by the process of *decomposition*, which resolves
bodies into their constituent parts, and also by our theory
of the atmosphere, by which its limits and operations are
determined.   But were we to abstract our attention from
the more general view of the subject, and confine our
inquiries to the German Ocean, the Baltic, the Mediterranean, the Red Sea, or to any other inland and circumscribed parts of the ocean, this difficulty seems to be
lessened.   Indeed, the probability is, and it is a pretty
generally received opinion, that a greater quantity of
water is actually admitted at the Straits of Gibraltar
and of Babelmandel than flows out of the Mediterranean
and Red Seas.   We consider water, therefore, as the
great *pabulum of nature*, which, as before noticed, enters
either simply or chemically into the constitution of all

bodies, and appears to be held, almost exclusively, in solution, in the formation and maintenance of the whole animal and vegetable kingdoms, and is found to exist largely in the composition of all mineral substances. The quantity of water, consequently, that is required, and is continually supplied from the ocean by the process of evaporation, both for the support and reanimation of nature, must be immense, and may of course be supposed permanently to absorb a very large proportion of the surplus waters of these circumscribed seas, while the remaining portion of surplus water, if not thus wholly accounted for, may be distributed over the general expanse of the ocean.

"But if we suppose with some, that in nature there is neither an excess nor diminution of the waters of the globe, and that the united and counterbalancing processes of evaporation, condensation, decomposition, and regeneration, so completely equalise each other, that the surplus waters, arising from the displacement of a portion of the solid surface of the globe, must again be wholly distributed and intermixed with the waters of the ocean, the portion of water remaining thus to be accounted for becomes more considerable, and, upon the great scale, must be permanently disposed of, independently of the process of evaporation.

"Another view has been suggested as applicable to the distribution of the surplus waters produced by the gradual filling up of the bed of the ocean. These waters, in place of being elevated in any sensible degree, may be naturally disposed to find their level in the great polar basins, or oblate portions of the surface of the globe which are known to exist next the poles. The oblate figure of the earth at the poles makes these imaginary points the nearest to the centre of the earth, and consequently, with regard to level, they are also the lowest. It therefore appears to follow, that any filling up of the bed of the sea near the equator, or at a distance from the poles, will have the effect of promoting the retiring of the surplus waters to the polar regions by their own gravity, while the centrifugal force occasioned by the earth's diurnal motion will prevent their being further removed from the earth's centre, without a corresponding elevation of the waters in the great polar basins.

" In this manner, such an accumulation of water may, at a former period of time, have taken place at the then poles of the globe, as to have altered the position of these points, and given rise to the Flood, or temporary general overflowing of the waters over the earth's surface, producing a change in the beds of the seas or oceans of former times. In this way may have been produced many of the phenomena observable in the crust of the earth, which are otherwise with much difficulty accounted for.

" Of what has now been advanced, regarding the waste of the land by the operations of the sea, it will be proper to notice that much consists with my own personal observation. The consequences of this process must be the deposition of débris, and a tendency to raise the bottom of the ocean and produce a proportional elevation of the water. With regard, however, to the distribution of the surplus waters that is produced, what I have now said is offered with much deference, in the hope that some one better qualified than myself will turn his attention to this curious subject."

In connection with this discussion I give the following interesting account of observations on the estuary of the Mersey :—

"Wasting Effects of the Sea on the Shore of Cheshire between the Rivers Mersey and Dee. (Read before the Wernerian Society, 8th March 1828.)

"On a former occasion I had the honour to make a few observations which appeared in the second volume of the Society's Memoirs regarding the encroachment of the sea upon the land generally. The present notice refers only to that portion of the coast which lies between the rivers Mersey and Dee, extending to about seven miles.

"To this quarter my attention, with that of Mr. Nimmo, Civil Engineer, had been professionally directed in the course of last month. In our preambulatory survey we were accompanied by Sir John Tobin and William Laird, Esq., of Liverpool, along the Cheshire shore and its connecting sandbanks between Wallasey Pool in the Mersey, and Dalpool in the river Dee.

"Within these estuaries the shores may be described as abrupt, consisting of red clay and marl, containing many land or boulder stones of the cubic contents of several tons, and very many of much smaller size, diminishing to coarse gravel. But the foreland or northern shore between these rivers, which I am now to notice, is

2 F

chiefly low ground, and to a great extent is under the
level of the highest tides.   The beach or ebb extends
from 300 to 400 yards seaward, and toward low water
mark exposes a section of red clay; but toward high
water it consists of bluish coloured marl, with peat or
moss overlaid by sand.   This beach, at about tide level,
presents a curious and highly interesting spectacle of the
remains of a *submarine forest*.   The numerous roots of
trees, which have not been washed away by the sea, or
carried off by the neighbouring inhabitants for firewood,
are in a very decayed state.   The trees seem to have
been cut off about two feet from the ground, after the
usual practice in felling timber, and the roots are seen
ramifying from their respective stumps in all directions,
and dipping towards the clay subsoil.   They seem to
have varied in size from eighteen inches to perhaps thirty
inches in diameter, and when cut with a knife appear to
be oak.   Several of the boles or trunks have also been
left upon the ground, and being partly immersed in the
sand and clay, are now in such a decomposed state that,
when dug into with a common spade, great numbers of
the shell fish called *Pholas candida*, measuring about
three fourths of an inch in length and two inches in
breadth, were found apparently in a healthy state.   These
proofs of the former state of this ebb or shore—now upwards of twenty feet under full tide—having been once
dry land to a considerable extent beyond the region of
these large forest trees were rendered still more evident
by the occurrence of large masses of greenstone, which, at
a former period, had been embedded in the firm ground

here, and especially on the shore within the river Dee.
It may further deserve notice that the inhabitants of this
district have a traditional rhyme expressive of the former
wooded state of this coast, where not a tree is now to be
seen, viz., "From Birkenhead to Helbre a squirrel
may hop from tree to tree;" that is, from the Dee to the
Mersey, now presenting a submarine forest.

" As these evidences of great changes upon the state and
former appearances of the land were highly interesting to
the party, and intimately connected with the professional
inquiries of myself and colleague, it seemed desirable, if
possible, to get them corroborated by oral testimony.
Sir John Tobin accordingly very obligingly took measures
for examining the oldest people in the neighbourhood as
to their recollection of the former state of these shores. In
particular, Thomas Barclay, aged ninety-three, "all but
two months," by profession a mason and measurer of
country work; Henry Youd, labourer, aged eighty-six;
and John Crooksan, labourer, aged eighty, were examined.
Barclay stated that he had been employed at the erection
of the Leasowe landward lighthouse in the year 1764;
that there were then two lighthouses near the shore, for
a leading direction to shipping through the proper channel
to Liverpool; and that the seaward light became unin-
habitable from its being surrounded by the sea. A new
light was then built upon Bidstone Hill, and the present
Leasowe Lighthouse, formerly the landward light which
he had assisted in building, became the sea light. He
could not condescend upon the distance between the two
original lights, but was certain that it must have been

several hundred yards; that he knows that in the course
of thirty years the shore of the Leasowe lost *by measurement* eleven Cheshire roods or eighty-eight yards; and
verily believes that, since he knew this shore, it has lost
upwards of half a mile of firm ground. To the correctness
of these statements the other two aged men gave ample
testimony, Henry Youd having also worked at the lighthouse.

" As to the present state of things, the party alluded
to were eye-witnesses of the tides on the 16th, 17th, and
18th of February 1828, having exhibited a very alarming
example of the encroachment of the sea upon the Leasowe
shore. At high water it came over the bank, and ran in
a stream of about half a mile in breadth surrounding the
lighthouse, and continued its course through the low
grounds toward Wallasey Pool on the Mersey, thereby
forming a new channel, and threatening to lay several
thousands of acres of rich arable and pasture lands into
the state of a permanent salt lake. The present Leasowe
Lighthouse, which, in 1764, was considered far above the
reach of the sea, upon the 17th of February last was thus
surrounded by salt water, and must soon be abandoned
unless some very extensive works be undertaken for the
defence of the beach, the whole of the interior lands of
the Leasowe being considerably under the level of high
water of spring tides.

" This coast, with its sandbanks in the offing, its submarine forest, and the evidence of living witnesses as to
the encroachment of the sea upon the firm ground, is
altogether highly interesting to the geological and scien-

tific inquirer. The remains of forests in the bed of the
ocean occur in several parts of the British coast, particularly off Lincoln, on the banks of the Tay near Flisk, at
Skail in the mainland of Orkney, and in other places
noticed in the Transactions of this Society, and are strong
proofs of the encroachments of the sea upon the land.
However difficult, therefore, it may be to reconcile the
varied appearances in nature regarding the sea having at
one time occupied a higher level than at present, yet its
encroachment as a general and almost universal principle
seems to be beyond doubt in the present day.

" Since I had last the honour of addressing the Society
on this subject, opportunities have been afforded me of
making many additional observations on the British
shores, and of personally extending these to almost every
port on the Continent between the Texel and the Garonne.
I have also, through the obliging communications of friends,
been enabled to extend my inquiries to other quarters of
the globe, and I am now prepared to state that, with a
few comparatively trifling exceptions, the sea appears to
be universally gaining upon the land, tending to confirm
the theory that débris arising from the general degrada-
tion of the land, being deposited in the bed of the minor
seas, is the cause of their present tendency to overflow
their banks."

DENSITY OF SALT AND FRESH WATER.

Mr. Stevenson's discovery that the salt water of the
ocean flows up the beds of rivers in a stream quite distinct
from the outflowing fresh water, was made in 1812, when

investigating a question regarding salmon fishings on the
Dee. It is described in the following extract from his
Report :—

" The reporter observed in the course of his survey that
the current of the river continued to flow towards the sea
with as much apparent velocity during flood as during
ebb tide, while the surface of the river rose and fell in a
regular manner with the waters of the ocean. He was
led from these observations to inquire more particularly
into this phenomenon, and he accordingly had an apparatus prepared under his directions at Aberdeen, which,
in the most satisfactory manner, showed the existence of
two distinct layers or strata of water ; the lower stratum
consisting of salt or sea water, and the upper one of the
fresh water of the river, which, from its specific gravity
being less, floated on the top during the whole of flood as
well as ebb tide. This apparatus consisted of a bottle or
glass jar, the mouth of which measured about two and a
half inches in diameter, and was carefully stopped with a
wooden plug, and luted with wax ; a hole about half an
inch in diameter was then bored in the plug, and to this
an iron peg was fitted. To prevent accident in the event
of the jar touching the bottom, it was coated with flannel.
The jar so prepared was fixed to a spar of timber, which
was graduated to feet and inches, for the convenience of
readily ascertaining the depth to which the instrument
was plunged, and from which the water was brought up.
A small cord was attached to the iron pin for the purpose
of drawing it, at pleasure, for the admission of the water.
When an experiment was made the bottle was plunged

into the water; by drawing the cord at any depth within the range of the rod to which it was attached, the iron peg was lifted or drawn, and the bottle was by this means filled with water. The peg was again dropped into its place, and the apparatus raised to the surface, containing a specimen of water of the quality at the depth to which it was plunged. In this manner the reporter ascertained that the salt or tidal water of the ocean flowed up the channel of the river Dee, and also up Footdee and Torryburn, in a distinct stratum next the bottom and under the fresh water of the river, which, owing to the specific gravity being less, floated upon it, continuing perfectly fresh, and flowing in its usual course towards the sea, the only change discoverable being in its level, which was raised by the salt water forcing its way under it. The tidal water so forced up continued salt; and when the specimens from the bottom, obtained in the manner described, were compared with those taken at the surface by means of the common hydrometer of the brewer (the only instrument to which the reporter had access at the time), the lower stratum was always found to possess the greater specific gravity due to salt over fresh water."

THE HYDROPHORE.

The instrument Mr. Stevenson then invented and used was that to which the term *hydrophore* has been applied. Figs. 18 and 19 show two forms of hydrophores made under his directions.

Fig. 18 is used for procuring specimens of water from moderate depths, drawn on a scale of one-tenth of the full

size. It consists of a tight tin cylinder, *a*, having a
conical valve in its top, *b*, which is represented in the
diagram as being raised for the admission of water.   The

valve is fixed *dead*, or immoveable, on
a rod working in guides, the one resting between two uprights of brass
above the cylinder, and the other in
its interior, as shown in faintly dotted
lines. The valve rod is by this means
caused to move in a truly vertical line,
and the valve attached to it consequently fills or closes the hole in the
top of the cylinder with greater accuracy than if its motion was undirected.
A graduated pole or rod of iron, *c*,
which in the diagram is shown broken

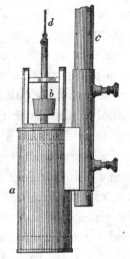

FIG. 18.

off, is attached to the instrument, its end being inserted
into the small tin cylinder at the side of the large water
cylinder, and there fixed by the clamp screws shown
in the diagram ; the bottom of the water cylinder may
be loaded with lead to any extent required, for the
purpose of causing the apparatus to sink ; but this, when
an iron rod is used for lowering it, is hardly necessary.
The spindle carrying the valve has an eye in its upper
extremity, to which a cord is attached for the purpose of
opening the valve when the water is to be admitted, and
on releasing the cord, it again closes by its own weight.
When the hydrophore is to be used, it is lowered to the
required depth by the pole which is fixed to its side, or,
if the depth be greater than the range of the pole, it is

loaded with weights, and let down by means of a rope so attached as to keep it in a vertical position. When the apparatus has been lowered as far as is required, the small cord is pulled, and the vessel is immediately filled with the water which is to be found at that depth. The cord being then thrown slack, the valve descends and closes the opening, and the instrument is slowly raised to the surface by means of the rod or rope, as the case may be, care being taken to preserve it in a vertical position.

The form of hydrophore represented in Fig. 19 is used in deep water, to which the small one just described is inapplicable. It consists of an egg-shaped vessel *a*, made of thick lead to give the apparatus weight, having two valves, *b* and *c*, one in the top and another in the bottom, both opening upwards; these valves (which are represented as open in the diagram) are, to insure more perfect fitting, fixed on separate spindles, which work in guides, in the same manner as in the instrument shown in Fig. 18. The valves, however, in this instrument are not opened by means of a cord, but by the impact of the projecting part, *d*, of the lower spindle on the bottom, when the hydrophore is sunk to that depth. By this means the lower valve is forced upwards, and the upper spindle (the lower extremity of which is made nearly to touch the upper extremity of the lower one, when the valves are shut) is at the same time forced up, carrying along with it the upper valve, which allows the air to escape, and the water rushing in fills the vessel. On

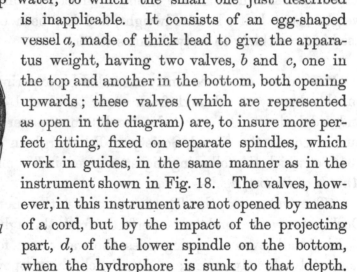

Fig. 19.

2 G

raising the instrument from the bottom, both valves again shut by their own weight, and that of the mass of lead, *d*, which forms part of the lower spindle. The mode of using this hydrophore is sufficiently obvious; it is lowered by means of a rope, made fast to a ring at the top, as shown in Fig. 19, until it strikes on the bottom, when the valves are opened in the manner described, and the vessel is filled; on raising it the valves close, and the vessel can be drawn to the surface without its contents being mixed with the superincumbent water through which it has to pass. This instrument, shown on a scale of one twentieth of full size, weighs about half a hundredweight, and has been easily used in from thirty to forty fathoms water.

Mr. Stevenson subsequently extended his experiments on the density of salt and fresh water to several firths and tidal rivers, and gave the results in a paper communicated to the Royal Society of Edinburgh in May 1817, of which the following digest is given in Thomson's *Annals of Philosophy* :[1]—

"The waters of the Thames opposite the London Dock gates were found to be perfectly fresh throughout; at Blackwall, even in spring tides, the water was found to be only slightly saline; at Woolwich the proportion of salt water increases, and so on to Gravesend. But the strata of salt and fresh water are less distinctly marked in the Thames than in any of those rivers on which Mr. Stevenson has hitherto had an opportunity of making his observations. But these inquiries are meant to be ex-

[1] Vol. x. p. 57.

tended to most of the principal rivers in the kingdom, when an account of the whole will be given.

"From the series of observations made at and below London Bridge, compared with the river as far up as Kew and Oxford, Mr. Stevenson is of opinion that the waters of the Thames seldom change, but are probably carried up and down with the turn of the alternate tides for an indefinite period, which, he is of opinion, may be one, if not the principal cause of what is termed the extreme softness of the waters of the Thames.

"Mr. Stevenson has made similar experiments on the rivers Forth and Tay, and at Loch Eil, where the Caledonian Canal joins the Western Sea. The aperture at Corran Ferry, for the tidal waters of that Loch, being small compared with the surface of Loch Eil, which forms the drainage of a great extent of country, it occurred to him that the waters of the surface must have less saline particles than the waters of the bottom. He accordingly lifted water from the surface at the anchorage off Fort William, and found it to be 1008·2; at the depth of 9 fathoms 1025·5; at the depth of 30 fathoms, in the central parts of the Loch, it was 1027·2; being the specific gravity of sea water."

The hydrophore, which was originally devised and used by Mr. Stevenson, in 1812, at Aberdeen, has now reached its height of excellence of construction and scientific importance in the famous ' Challenger' Expedition.

CHAPTER XVII.

EXTRACTS FROM EARLY REPORTS.

Wide range of subjects on which Mr. Stevenson gave advice—Reports on ruins of Aberbrothock Abbey—St. Magnus Cathedral, and Earl's Palace, Kirkwall—St. Andrews Cathedral—Montrose Church Spire—Melville Monument, Edinburgh—Lipping of joints of masonry with cement—Provision for flood waters in bridges—Hydraulic mortar—Protection of foreshores—Cycloidal sea wall—Checking drift sand—Night signal lamps—Cause of heavy seas in Irish Channel—Sea routes across Irish Channel—Build of ships—Prospective increase of population—Tidal scour—Unscrewing of bolts by the waves—Cement Rubble cofferdams—Buoyage system—Observations on fog signals—Regulations for steam vessels—Notes on shipwrecks.

JUDGING from Smeaton's well known "Reports," to which all have access, we may conclude that the "professional advice" given by early Engineers was very generally accompanied by a fuller and less reserved discussion of opinion than is to be met with in the brief and technical Engineering reports of the present day. In early times, Engineers did not hesitate to express themselves freely on physics, æsthetics, or commerce, provided their views had a collateral bearing on the subject under discussion, and this often added to the interest of their reports.

These early Engineers were also consulted on a much wider range of subjects than the Engineers of modern times. We know that the larger requirements of modern Engineering demand that its practice should be classified under distinct branches, such as harbours, navigations, water works, gas works, lighthouses, or railways, not to mention electrical and sanitary engineering, and other branches of modern growth, all of which cannot possibly be advantageously practised by any one member of the profession; for no one mind can grasp the theoretical knowledge, and no one life can compass the practical experience, to enable a man to attain eminence in all these departments of modern Engineering.

A biographical sketch of Mr. Stevenson's professional life would, it seems to me, be incomplete if it did not convey to the reader some notion, however general, of the wide range of subjects brought under his notice, in these early times, and of his comprehensive and suggestive mode of treating every case on which he was professionally consulted. This object would be only imperfectly attained were I to restrict my reference to his reports to the examples given in the preceding chapters; for I have found in his numerous writings casual notices of a miscellaneous and fragmentary character, many of which seem to me to be interesting to the profession, and worthy of preservation, and I propose, in this chapter, to give a few of these extracts, without order of subject or date; and I think they will justify my remark as to the great variety and fulness of treatment to be found in the reports of early Engineers.

It appears, for example, that Mr. Stevenson was often called to advise on matters which were more related to architecture than engineering. Of this nature was his tour of inspection to the jails of England, in company with Sir William Rae, the Sheriff of Edinburgh, in 1813, referred to in a former chapter.

ABERBROTHOCK RUINS.

In like manner he inspected Aberbrothock Abbey, with Sir Walter Scott and the Sheriff of Forfar, in 1809, to advise as to preserving the ruins, some of the turrets being in imminent danger of falling; and after procuring a survey of the whole building he prepared a report, with plans and specification, which were submitted to the Barons of Exchequer, and the work was thereafter carried out under his direction.

ST. MAGNUS CATHEDRAL AND EARL'S PALACE.

He also reported in a similar way to the Sheriff of Orkney with reference to the repairs of the Earl's Palace at Kirkwall, estimated at £500, and on certain alterations at the Cathedral of St. Magnus.

ST. ANDREWS CATHEDRAL AND MONTROSE SPIRE.

With a similar object in view he inspected and reported on the Cathedral of St. Andrews, and the steeple of the Church of Montrose, which was thought to be in danger, and the result of that inquiry was the present beautiful spire, built from the designs of James Gillespie Graham.

MELVILLE MONUMENT.

He was also associated with Mr. Burn in the Melville Monument of Edinburgh,—the preparation of the foundation, the rubble work for the tower, and the scaffolding and tackling for raising the statue were carried out under Mr. Stevenson's direction ; the whole architectural design being due to Mr. Burn alone.

LIPPING OF JOINTS OF MASONRY WITH CEMENT.

The well known practice of what is termed "lipping" with cement the mortar joints of masonry exposed to the wash of water is described by him as new in his report to the Trustees of Marykirk Bridge, of 16th July 1812, where he says :—

"Upon carefully examining the face joints of the masonry of the south pier under water line, some of these were found not to be so full of mortar as could have been wished, and although Mr. Logan (the inspector of works) had taken the precaution to cause the joints to be covered with clay to preserve them from the effects of the water, yet this had not altogether answered the purpose, and hence the reporter recommended to the meeting of the 8th current *to provide a few casks of Parker's Roman Cement, to be laid to the breadth of three or four inches upon the bed and end joints under the low water mark of the remaining piers.*"

PROVISION FOR FLOOD WATER IN BRIDGES.

In determining the waterway of his bridges, Mr. Stevenson invariably provided for prospective increase of flooding due to agricultural improvements, as stated in the following extract from a report made in 1811 :—

"To preserve an ample waterway the north abutment is placed about twelve feet from the edge of the river,

leaving a sufficient passage for the water in floods. A
less waterway might perhaps have answered the purpose,
but as the valleys through which the North Esk passes
may come to be meliorated by drainage, and especially
those districts of country on each side of the feeders which
join the river, the facility with which the surface water
may then escape must greatly increase the floods, and
although their duration will be shorter, yet their rise
must be proportionally higher."

HYDRAULIC MORTAR.

The following remarks on hydraulic mortar, made in
1811 to the Commissioners of Montrose Bridge, are interesting as showing the detail which he brought to
bear on all his works :—

"The best mortar for water work is a mixture of
Pozzolano earth with lime and sand, but the late interrupted state of commercial intercourse with the Mediterranean has for years past rendered Pozzolano so scarce
an article as hardly to be procured on any terms. Your
reporter has therefore been induced to make various
experiments with preparations of lime and Roman cement,
and finds that a mixture may be made which will set
under water and answer every purpose. For this
mortar the lime ought to be well burned, and put into
casks when drawn from the kiln. It should be brought
to the work as recently after being burnt as possible.
This will be most readily attained by taking the lime
from Boddam kilns. English lime is in general stronger
and cleaner, but some of it brought for the purpose of

agriculture is not so suitable for buildings as Lord Elgin's lime. These limes, however, cannot be had very newly burnt, and it will be preferable to take lime from some of the kilns in the neighbourhood which are of good character. When brought to the bridge the lime should be kept under cover, opening only one barrel at a time; the shells must be pounded to a state of powder, and immediately before mixing it with the other ingredients it will be proper to sprinkle a little water upon it to dissolve any gritty particles that may remain amongst it.

" The sand for this work, though fine, must nevertheless be sharp; it must also be passed through a sieve, and cleaned of all impurities by washing, if found necessary. For ramming the joints and pointing under water, let equal parts of lime in its powdered state and of Roman cement be used, with one fourth part of prepared sand, but for the upper works the quantity of Roman cement in the mortar may be reduced to one third part.

" The mortar must be mixed in small quantities and quickly beaten up into a consistency suitable for the work. All white specks, which are apt to swell and spoil the joints, must be carefully rejected from the mortar."

PROTECTION OF FORESHORES.

Some suggestive remarks on the protection of foreshores, made in 1812, in a report to Lord Rosebery, on his Lordship's property at Barnbougle Castle on the Firth of Forth, are given in the following terms:—

" If the operation of the waters of the ocean be

2 H

attended to in the formation of the shores, some useful
hints may be gained.   These shores will be found to be
so many inclined planes, varying in declivity according
to the tenacity of the matter of which they are composed.
Hence it is that the minute grains of sand and the light
sea shell become a lasting barrier against the rapid river
current and the tumultuous ocean, while the erect sea
wall is levelled with the ground.   For the truth of this
it were needless to refer to the works of nature in
different quarters of the world, or in distant parts of this
country; it is only necessary to examine the shores on
each side of Barnbougle Castle, where the beautiful
beach, consisting of sand and shells, between the Cockle
Burn and the sea, forms a complete defence to the low
grounds behind it, while to the northward of the castle
the massive wall is in danger of being completely thrown
down.   Without waiting to inquire into the causes which
regulate these appearances, it will be more consonant
to the business of this report to point out how their
simple forms may be imitated and turned to advantage."

CYCLOIDAL SEAWALL.

In reporting on the defence of the lands of Trinity,
on the Firth of Forth, Mr. Stevenson recommended the
adoption of a cycloidal talus wall, which was executed
under his direction in 1821 :—

"In giving an opinion relative to the best mode of
defending and preserving this property, the reporter
observes that it fortunately happens that the beach is
pretty closely covered with large boulder stones, which
now form a kind of *chevaux de frise* in breaking the force

of the sea, and making it fall more gently towards high
water mark. Were it not that these stones are proposed
to be employed in the erection of a more effectual barrier
against the waves, the reporter would not fail to dis-
approve of their removal for any other purpose.

"The reporter proposes that a *Talus wall* or bulwark
should be built of these boulder stones, roughly dressed
and laid so as to form a cycloidal curve in the central
part, as nearly as may be, as represented in the section
with its tangents (Fig. 20). The properties of the

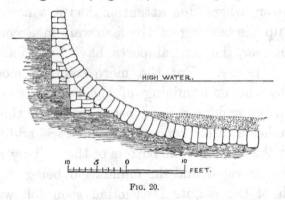

HIGH WATER.

10 5 0   10
|‖‖‖‖‖‖‖‖|_____| FEET.

Fig. 20.

cycloid as applicable to a sea wall in an exposed situation
are very important. In particular, if compared with any
other curve, in the same vertical line and down through
the same points, it will be found of swiftest descent
under similar circumstances, therefore the water in its
rise must be proportionally retarded. The lower tangent
to the curve alluded to also forms a wall towards low
water, best adapted for admitting the sea to flow gently
over it, while that connected with the upper extremity
of the cycloidal part, tending towards the perpendicular,
brings gravity into action against the rise of the waves.

The practical execution of a wall upon this construction is simple, while the aggregate quantity of materials is less than for any of the curves of the conic sections of similar extent, and it seems upon the whole to be peculiarly applicable for the defence of the sea beach in question.

"If we examine the numerous works of this kind erected for similar purposes along this coast, we shall find that the general process or action of the waves is to undermine the seaward courses of the walls. In some cases, however, where due attention has not been paid to making up the backing of the face wall in a compact and firm manner, the central parts have been found to sink and give way. But the more common mode of failure is by the undermining of the seaward courses, arising from too sudden a slope being given to the face wall, which has a direct tendency to produce additional agitation in the waters at the bottom of the wall, by which the beach is excavated, and the foundation, being exposed to the wash of the sea, its destruction soon follows. If we attend to the distribution which nature makes of the matters composing a sea beach, unless where special local causes occur, we find them laid with a very gradual descent towards low water mark. The sands of Portobello, in this neighbourhood, form a striking example of this. Here small quartzose grains mixed with light sea shells prove, in their effects, a more effectual barrier against the overwhelming force of the waves than perpendicular and massive walls of masonry."

CHECKING DRIFT SAND.

Mr. Stevenson recommended Lord Palmerston to introduce the *Pinus maritima major*, as a check for sand drift, on his estate of Mullaghmore, in the following report, dated 21st July 1835 :—

" During the reporter's visit to Mullaghmore, his advice was also asked regarding the operations at present going on for the improvement of the land. He had then much satisfaction in viewing the interesting improvements of reclaiming bog lands, and checking the inroads of the sand flood or drift, by planting 'bent' grass upon the shores of this estate. The system of dibbling the bent grass, pursued by Mr. Lynch, is in the best style which the reporter has anywhere met with ; and he has been so impressed with the national importance of this scheme, from the success already experienced at Mullaghmore, that he has already taken the opportunity of recommending this system as applicable to the entrance of Ballyshannon, and in other quarters, particularly to the Highland and Agricultural Society of Scotland.

" The question chiefly submitted to the consideration of the reporter, in regard to these operations, was the best mode of defending the margin of the bent grass towards the sea. For such purposes, buildings or fences of any kind are not only expensive in their formation, but are also in constant need of repair. Mr. Lynch seems so much at home in all planting operations that the reporter begs simply to bring under your Lordship's notice the French mode of planting a species of fir (*Pinus maritima major*), which was originally suggested to the Government by the late M. Bremonteuil, *Ingénieur des Ponts et Chaussées.* This system has been extensively tried along the stormy shores of the Bay of Biscay, particularly in the district of Grave, at the entrance of the Garonne, where the arid and sterile sands have been covered with extensive forests, which thrive quite close to the water's edge. From the climate and exposure of the shores at Mullaghmore, the reporter has no doubt of the success of similar plantations in arresting the progress of the sand flood. It is believed that Mr. Lawson, seedsman to the Highland and Agricultural Society of Scotland, is taking measures to import the seeds of the *Pinus maritima major,* with a view to trying it on some of the exposed sandy districts of Scotland."

From the following extract of a letter from Mr. Kincaid of Dublin, who was Lord Palmerston's Commissioner, it is interesting to know that the experiment was entirely satisfactory, proving that the *Pinus maritima major* is well adapted to the climate of the coasts of the British Isles :—

"The Mullaghmore plantations extend to about 200 acres. About eighty of these were planted twenty-five years ago. Some of the trees are thirty feet in height, and vary from that height to about twenty or twenty-five feet. The remainder were planted ten years ago, and are making fair progress. All the pine plantations from opposite Newtown Cliffony to Mullaghmore are in a most healthy condition, the trees making growths of from twelve to twenty inches each year. The storms have no bad effect on the south side of the great sand hill, but on its summit, and towards the west side, the spray and gales of the Atlantic will not allow the young trees to make any progress."

NIGHT SIGNAL LAMPS.

In a report to the Trustees for improving the Queensferry passage, made in 1811, Mr. Stevenson proposes a set of signals as described in the following extract, his proposal being, in fact, the signal now in use on all British railways :—

" Upon the supposition of its being the intention of this Honourable Trust to have an establishment on the south side of the Firth similar to that which is now proposed for the north side, the reporter takes the liberty of observing that much advantage, as the Trustees know, might be derived by the public from a few simple and well appointed signals, both for night and day.

" Those intended for the day may be constructed upon a modified scale, after the common telegraphic method ; while the night signals can be rendered extremely simple and effective by interposing at pleasure between the observer and the reflector a shade of coloured glass. By connecting these partial obscurations of colouring the light with an index that shall be understood on both sides of the passage, orders may be communicated in a very expeditious manner."

CAUSE OF HEAVY SEAS IN IRISH CHANNEL.

In a report to the Right Honourable Viscount Cathcart, Commander of His Majesty's Forces, made on Portpatrick harbour in 1812, he gives the following explanation of the well-known rough sea between Portpatrick and Donaghadee :—

"In describing the harbour of Portpatrick, it may be noticed that although the coast on which it is situated is not directly exposed to the Atlantic Ocean, yet the opposing tides of the north and south channels meet there and separate to flow up the Clyde and Solway Firths, which, independent of storms, must occasion a very considerable commotion in the waters of the channel between Portpatrick and Donaghadee.

"Accordingly we find that the sea has made a great impression upon the coast of Wigtonshire; and though the shores between Loch Ryan and the Bay of Glenluce consist chiefly of whinstone (the greenstone of mineralogists), which is one of the most indestructible rocks we have, yet the figure of the coast is indented with many small cuts or creeks, and rocks are all along the shore found jutting into the sea. At the head of one of these creeks, which is about a hundred fathoms in length, and thirty fathoms in breadth, the harbour of Portpatrick is situated between two insulated rocks, upon one of which the piers are built, the harbour being formed by an excavation, chiefly in the solid rock."

SEA ROUTES ACROSS IRISH CHANNEL.

In the same report he states the relative advantages of various routes of communication across the Irish Channel:—

"A further extension of the intercourse between Scotland and Ireland could be made with much advantage to both by a regular establishment of packets between Ardrossan, Troon, or Dunure in Ayrshire, and Larne in the county of Antrim. Between the two last places, viz., Dunure and Larne, the distance would only be about sixty miles, being ten miles shorter, and unquestionably much safer, than the passage from Holyhead to Dublin.

"Under all the views of this subject, from the greater contiguity of Portpatrick and Donaghadee than of Lochs Ryan and Larne, and the former places having more immediate access to the open sea than the latter, and also from the intercourse being now fully organised by long establishment, it were perhaps better, even at a much greater expense, to continue the present system than to change it. Portpatrick harbour may be rendered incomparably better by the plan now proposed, and Donaghadee is also capable and stands much in want of improvement, by an extension of its piers and the erection of a permanent light to direct the packets into the harbour under night."

BUILD OF SHIPS.

In reporting to the Royal Burgh of Dundee as to the improvement of the harbour in 1814, Mr. Stevenson takes occasion to introduce one of those collateral questions to which I have referred :—

"It is curious to observe the changes and to trace the progressive improvements which have taken place in the form and *build* of ships. When we contrast those of early navigators with ships of modern times, among the many alterations, none seems more striking than the difference of their depth. The draught of water that was required for a ship of 300 tons burden would hardly be found enough to float a modern built vessel of 100 tons.

2 I

This alteration in the construction of ships, which is mainly calculated to improve their sailing, by giving them a better hold of the water, seems gradually to have advanced, as the mariner became more adventurous in his voyages; and is only now restrained by certain considerations of convenience, of which the most prominent is the want of a sufficient depth of water in the havens and harbours on the coast for their reception, —a circumstance which arises partly from the natural position of harbours, but is chiefly owing to the difficulties and expense attending the necessary engineering operations, which increase enormously with the depth of water. Hence it is that many of the ancient seaport towns of this country, which at one time possessed an extensive trade, have, from neglecting their harbours, sunk into a state of insignificancy; while others, by proper exertions in this respect, have, under the most inauspicious circumstances, attained to great commercial importance."

PROSPECTIVE INCREASE OF POPULATION.

Another case of the same kind occurs in his report on the harbour of North Berwick, made in 1812 :—

"Before closing this report it may be noticed that North Berwick has considerable advantages, which if acted upon would infallibly lead to the rapid improvement of the town and neighbourhood. Situated upon an extensive flat which skirts along the high land of North Berwick Law, on a beautiful sandy bay, which is intersected by the street leading to the harbour, few towns will more easily admit of elegant extension or are better calculated for becoming a sea-bathing retreat."

It has now the well-known reputation of being the best frequented watering place on the east coast of Scotland.

TIDAL SCOUR.

In the report, of 1814, on Dundee we find the following remarks on tidal scour :—

"To put this matter in a clearer point of view, let us see what nature does upon the great scale, as for example in the extensive basin forming the Firth of Tay. We there find that in consequence of the rapidity of the current at the narrow passage in the neighbourhood of Broughty Castle, which may be viewed as the *scouring aperture* of the basin of the Tay, the water is from forty to eighty feet in depth, and moves with a velocity which carries a great quantity of sandy particles along with it. But no sooner are the waters of this current allowed to spread and cover the basin of the Tay, than the velocity ceases, and the foreign matters fall to the bottom and form the various sandbanks which appear at low water. In a similar way the deposition of silt and earthy particles brought down the river in speats is accounted for. Now, this view of the case is equally applicable to the harbour of Dundee, for so long as the water preserves the velocity it acquires in the *scouring apertures* or arches in the quays, it carries all its foreign matters along with it; but the moment it is allowed to expand over the extent of the harbour the deposition of these earthy particles begins.

And in every case the well-known law in hydraulics holds good, that the *scouring effect* of a fluid is in the ratio of the square of the velocity."

UNSCREWING OF BOLTS.

The following observations made in 1807 on the action of the waves in unscrewing bolts, are interesting :—

"The unlocking of screws, where *washers* had been introduced as a security was rather unexpected, and the writer took an opportunity of conversing with his much respected friend Professor Playfair regarding this circumstance. The Professor observed, that he had experienced some inconvenience of this kind from the unlocking of almost all the screws of a telescope which had been sent to him from London by the mail coach. Indeed, from the spiral form of the screw, which is, in fact, an inclined plane, Mr. Playfair readily accounted for such an occurrence, and, when reflected upon, it seems to be an effect rather to be looked for, and is a reason why riveting the point of a bolt in preference to screwing it should generally be resorted to, where much motion is to be apprehended."

CEMENT RUBBLE COFFERDAMS.

I give his description of the cement rubble cofferdams, first used in 1808, at the erection of the Bell Rock Lighthouse :—

"At seven o'clock this morning, the tide proving more favourable, the artificers began to work. At nine o'clock the rock was again overflowed, and the boats returned to the tender after two hours' work. Part of the operations of this morning's tide consisted in building up the crevices and inequalities of the rock round the margin of the foundation with Pozzolano mortar and the chips produced from excavation, with the view to dam out the water. These little walls varied from six to eighteen inches in height; a small sluice or aperture being formed in one of them, by which the water, during ebb tide, was allowed to drain off.

"It formed part of the writer's original design to erect a cast iron cofferdam of about five feet in height round the site of the building; but the surface of the rock was so irregular that the difficulty of tightening it, and also of emptying the contained water, so as to get the benefit of it during ebb tide, would have been so great, that taking these circumstances into account, together with the loss of time which would attend the erection of such a preparatory work, the idea of a cofferdam was laid aside, soon after entering upon the actual execution of the work."

BUOYAGE SYSTEM.

In his report on the Forth Navigation, made to the Magistrates of Stirling in 1828, Mr. Stevenson proposed a system of buoyage, which has since been adopted by the several Lighthouse Boards of the United Kingdom:—

"The channels proposed to be cleared through the different fords are coloured red on the Plan, in reference to the sectional line. For the use and guidance of river pilots, buoys and perches or beacons are likewise intended to be placed in the positions shown in the Plan; those coloured red are to be taken on the starboard, and those coloured black upon the larboard side, in going up the river; and the whole are to be so placed in connection with the clearing and deepening of the fords as to be approached with safety."

OBSERVATIONS ON FOG SIGNALS.

At a very early period Mr. Stevenson's attention was directed to the dangers of fog at sea, and the best means of providing an effective fog signal for the mariner, and so long ago as 1808 he had come to the conclusion that the best signal adapted for the purpose was the sustained sound of a horn, which, as is well known, has within the last few years been so much employed in the fog signals which are now being established at many of the light-house stations in this and foreign countries. The following extracts give an idea of the difficulties he encountered, and his views on the subject :—

"The boats landed this evening (23d June 1808), when the artificers had again two hours' work. The weather still continuing very thick and foggy, more diffi-culty was experienced in getting on board of the vessels to-night than had occurred on any previous occasion, owing to a light breeze of wind which carried the sound of the bell, and the other signals made on board of the vessels, away from the rock. Having fortunately made out the position of the sloop "Smeaton," at the north-east buoy, to which we were much assisted by the barking of the ship's dog, we parted with the Smeaton's boat, when the boats of the tender took a fresh departure for that vessel, which lay about half a mile to the south westward. Yet such is the very deceiving state of the tides that although there was a small binnacle and compass in the landing master's boat, we had nevertheless passed the 'Sir Joseph' a good way, when fortunately one of the sailors caught the sound of a blowing horn. The only fire-arms on board were a pair of swivels of one inch calibre; but it is quite surprising how much the sound is lost in foggy weather, as the report was heard but at a very short distance. The sound from the explosion of gunpowder is so instantaneous that the effect of the small guns was not so good as either the blowing of a horn or the tolling of a bell, which afforded a more constant and steady direction for the pilot. It may here be noticed that larger guns would have answered better, but these must have induced the keeping of a greater stock of gunpowder, which in a service of this kind might have been attended with risk. A better signal would have been a bugle horn, the tremulous sound of which produces a more powerful effect in fog than the less sonorous and more sudden report of ordnance."

And again he says :—

"In the course of this morning's work two or three apparently distant peals of thunder were heard, and the atmosphere suddenly became thick and foggy. But as the Smeaton, our present tender, was moored at no great distance from the rock, the crew on board continued blowing a horn, and occasionally fired a musket, so that the boats got to the ship without difficulty. The occurrence of thick weather, however, became a serious consideration in looking forward to the necessary change of quarters to the Pharos, distant about one mile from the rock, instead of a few hundred yards, as in the case of the Smeaton.

"The weather towards the evening became thick and foggy, and there was hardly a breath of wind to ruffle the surface of the water; had it not therefore been the noise from the anvils of the smiths, who had been left on the beacon throughout the day, which afforded a guide for the boats, a landing could not have been attempted this evening, especially with so large a company of artificers. This circumstance confirmed the writer's opinion with regard to the propriety of connecting large bells to be rung with machinery in the lighthouse, to be tolled day and night during the continuance of foggy weather, by which the mariner may be forewarned of too near an approach to the rock, while every distant object is obscured in the mist."

Following out this subject, Mr. Stevenson caused observations to be made at the Calf of Man—a small island at the south of the Isle of Man, and separated from the main island by a narrow "sound." The place is noted for its fogs, on which Mr. Stevenson says :—

"I sent Mr. Macurich, a shipmaster in the lighthouse service, to the Calf of Man, with directions to reside there, and make monthly returns of the state of the weather, agreeably to a printed form. During his stay of seven months, it appears upon the whole that the fog rested only twice upon the highest land of the Calf Island, while it cleared partially below. On one of these occasions I was on board of the lighthouse yacht, then at anchor off the island, when the fog was for a time general; and as the weather became clear, I observed that it first disappeared upon the lower parts of the island, and that in half an hour the whole of the Calf was seen. In the monthly returns made by Mr. Macurich, the Calf island is represented as often perfectly free of fog, while the higher parts of the opposite mainland of the Isle of Man were hid in mist. To account for this, it may be noticed that the mass of matter in the Calf Island is much less, and the land is also much lower than in the main island. Part of this effect may also be ascribed to the rapidity of the tides, which create a current of wind, particularly in the narrow channel between the main and Calf islands, which have a direct tendency to clear away the fog, as I have observed at the Skerries in the Pentland Firth, and in similar situations on different parts of the coast, where rapid currents prevail."

These extracts are given to show the attention Mr. Stevenson gave to the subject of fogs, which, as already noticed, led him to recommend the horn, the instrument now so much used in giving signals to the mariner.

Akin to this may be mentioned his expression of regret that no means existed for determining the force of the wind, as noticed in the following paragraph :—

"We cannot enough regret the want of an efficient anemometer, or instrument for measuring the force of the wind. Indeed, we hardly know any desideratum of more universal interest, for, notwithstanding the labours of

2 K

Lind and others on this subject, from the want of a proper scale we are still groping in the dark with the use of such indefinite terms as 'light airs inclining to calm,' 'fresh breezes,' 'fresh gales,' 'hard gales,' and 'very hard gales;' for it rarely happens that the sailor will admit the term 'storm' into his nomenclature."

REGULATIONS FOR STEAM VESSELS.

The loss of the 'Comet' steamer by collision on the Clyde, in 1825, led the Lord Advocate to entertain the idea of introducing a Bill for the regulation of steamers, and to issue a circular in the following terms, of which Mr. Stevenson received a copy :—

"EDINBURGH, *4th Feby.* 1826.

"I annex a copy of the heads of such a Bill as, in my opinion, may be calculated to afford sufficient security to steamboats, and thereby alike promote the interests of the owners of such vessels and that of the public. I feel noways wedded to any of the proposed provisions, and am anxious to submit them to the consideration of the better informed on such subjects, so as to obtain suggestions either as to the additions or amendments which the Bill may be fitted to receive.

"In directing your attention to this important subject, I need hardly remind you that in our endeavours to render such vessels perfectly secure in so far as respects the passengers, we must not lose sight of the interest of the owners, or attempt to clog the trade with unnecessarily embarrassing regulations. Such restrictions are seldom enforced, and, if they should receive effect, might lead to such harassing consequences as would injure this useful description of property, and thereby to a certain extent deprive the public of the great benefit which is now derived from the use of vessels navigated by steam. —I have the honour to be your most obedient servant,

"WM. RAE."

The only account I can find of Mr. Stevenson's views on this important subject is contained in the following extract from a letter, dated 3d November 1825, to Captain Foulerton, one of the Wardens of the Trinity House, with whom he appears to have had much correspondence, in which he explains views which are very much in accordance with the regulations for steamers now issued by the Board of Trade. His letter says :—

"We lately had a melancholy accident, as you would see, by the running down of the 'Comet' steam packet, by which, it is believed, that about seventy people lost their lives. The Lord Advocate attended himself at the taking of the precognition, and is, I believe, to bring some of the parties to trial. He has also in view some regulations by an Act on this new and important subject.

"From my seeming marine habits his Lordship has desired me to state what occurs on the subject of lights. If we need this on the Forth and Clyde, you must be in a worse state in the Thames. I have no doubt you had this under the notice of your House. I think there should be two lights, one in each bow, but under deck, in order to keep the lights *entirely* out of the view of those on deck.   I am not for interfering with their head sails.   I would have them licensed like stage coaches, and placed under the inspection of an officer of the navy, not below the rank of a lieutenant.   Six or eight officers might do the duty for the whole United Kingdom for a time."

The accident seems to have led to a further investigation into the general question of the saving of life in cases of shipwreck on the coasts of Scotland; and on this subject Mr. Stevenson made the following replies to the queries submitted to him by the authorities :—

"QUERY.—Are shipwrecks frequent on the coasts of Scotland and its islands?"

"Wrecks between the Firths of Forth and Moray are more frequent than on any other part of the coast of Scotland.   This may probably be accounted for by the great number of vessels passing and repassing along that coast. In the month of December 1799, a strong gale from the south-east occasioned serious disasters on these shores, when upwards of seventy sail were wrecked on the eastern coast of Scotland, and many of their crews perished.   This lamentable catastrophe was the means of causing lifeboats upon Greathead's plan to be fitted out at St. Andrews, Arbroath, Montrose, Aberdeen, Peterhead, and other places, which have been found highly useful in saving the lives of mariners.   This gale was also the immediate cause of the erection of the Bell Rock Lighthouse, which may be said almost entirely to have prevented shipwreck, so frequent in St. Andrews Bay and the entrance of the Firth of Forth in general.

"From the Moray Firth along the shores of the mainland to the entrance of the Firth of Clyde, wrecks cannot be said to be very frequent, although the navigation is rather difficult; but the safety of shipping on this coast depends upon the great number of excellent natural bays and harbours upon it.

"In the Orkney and Shetland Islands few seasons pass without wrecks occurring. On the Lewis and Western Hebrides shipwrecks frequently occur."

"QUERY.—Are the coasts of Scotland in general well provided with the means of giving assistance in case of shipwreck, or are they deficient in such provision?"

"The coast of Scotland is provided with no other means of saving the crews of vessels than the assistance they accidentally meet with from the inhabitants along shore. The only lifeboats established are those at the ports already mentioned.

"If Captain Manby's apparatus was generally known and applied upon the coast, it would be found highly beneficial."

"QUERY.—Are any instances remembered of total shipwrecks where lives lost might have been saved by the lifeboat or by Captain Manby's apparatus, at the distance of 350 or 400 yards off the coast?"

"In the year 1813 the 'Oscar,' Greenland ship of Aberdeen, Captain Innes, went ashore upon Girdleness, at the entrance of Aberdeen Harbour. There were on board fifty-four persons, of whom only two were saved, by dropping from the bowsprit end. The ship was very near the shore. She broke up about twenty minutes after she struck, and I have no doubt that, if an active person had been on the spot with Captain Manby's apparatus, the greater part of the crew of this ship might have been saved.

"In the winter of 1824 the 'Deveron' of Aberdeen, Captain Scott, went ashore upon the sands three miles north of Aberdeen in a gale at south-east. She was only about 300 yards from the shore, and here the whole crew must have perished had it not been for the prompt use of Captain Manby's apparatus.

"Every one who has seen this apparatus must have admired its simplicity and effect. It is however difficult to see how its application can be very generally introduced so as to be useful along the whole extent of chequered coasts of the British dominions. Certainly at all principal ports it would naturally be expected that both this and the lifeboat would be provided.

"A time seems to be approaching when the coast will be much more complete in all such provision from the hands of the humane for the safety of the mariner. We also hail with pleasure the extending efforts of the respective Lighthouse Boards on the coasts of England, Scotland, and Ireland, as a certain means of adding to the security of that useful body of men, as well as to the facilities of her enterprising merchants. Nor can we withhold the notice of the effect of the operations of the Scots Board in this respect. At the entrance to the Firth of Forth, prior to the erection of the Bell Rock

Lighthouse, few winters passed without some disastrous shipwreck.

"Even after the completion of this arduous undertaking, until the beacon was erected on the Carr Rock, off Fifeness, the fisherman's observation was—'The Carr has always her wreck : if she misses one year, she is sure to have two the next.' But since the erection of this beacon in 1820 till this date (1825), not a single wreck has happened on this part of the coast."

CHAPTER XVIII.

RETROSPECT OF MR. STEVENSON'S LIFE.

THE unconnected sketches which form this Memoir extend over a period of about forty years. They have, as already stated, been selected from among a large mass of documents, in order to convey to the reader, not only some idea of the great variety of subjects Mr. Stevenson was called on to consider, but also to show his happy power of dealing with engineering questions in the several aspects under which they were presented to him. In perusing them, the reader can hardly have failed to remark in how many instances the views Mr. Stevenson expressed were forecasts either of great fundamental social changes, such as the substitution of the railway for the road, or of smaller though important matters of detail, as, for example, the signal lights of our railways and steamers, without which the "night traffic"—so popular a feature of modern travelling—could not possibly be conducted. These and many other instances must have satisfied the professional reader that *foresight* and *originality* were remarkable features of Mr. Stevenson's character.

In the department of Lighthouses, he had ex-

periences which, it may be safely said, none of his com-
peers possessed, and I think it will be admitted that
in his general practice he displayed powers of observation of a high order. Acting as he did with Rennie,
Telford, Nimmo, and afterwards with Walker, George
Rennie, and Cubitt, with all of whom he ever remained
in friendly intercourse, his experience was both large and
varied, and the whole of his practice as an Engineer
was distinguished by full preliminary investigation of
his subject—great caution in forming his conclusions
—elaborate preparation of his reports and designs, and,
as specially called forth at the Bell Rock Lighthouse,
masterly skill, indomitable energy, and unwavering fortitude in carrying his designs into execution..

My father was elected a Fellow of the Royal Society
of Edinburgh in 1815, and soon after joined the Anti-
quarian and Wernerian Natural History Societies, taking
an active part at their meetings and communicating
papers to their proceedings. He was a Fellow of the
Geological and Astronomical Societies of London, a
Member of the Smeatonian Society, and of the Institution of Civil Engineers.

He was also one of the original promoters of the
Astronomical Institution, out of which has grown the
present establishment of the Royal Observatory of
Edinburgh, and the following account of the early origin
of the Institution was drawn up some years before Mr.
Stevenson's death at the request of Professor Piazzi
Smyth, the Astronomer-Royal of Scotland :—

"There was a young man named Kerr—an optician—in Edinburgh, who, on commencing business, brought about the formation of a Club, somewhat like a Book Club, for procuring philosophical instruments for the use of its members. These were more particularly optical instruments and theodolites, etc., for surveyors, which were also to have been lent out for hire. I think the subscription was a guinea. The meetings were, perhaps, monthly; they were held in the office of Mr. James Ogilvy, Accountant, Parliament Square.

"I attended two, or perhaps three, meetings in the year. The Club was formed before I was invited to become a member. At the first meeting I found present Mr. James Bonar, treasurer of the Royal Society; Mr. Christison, mathematician; Mr. Brown, bookseller, opposite the college; Mr. Ogilvy, and Mr. Kerr.

"After attending one or two meetings of this very modest Society for the advancement of science, Mr. Bonar and I had some conversation upon its prospects, and the difficulties attending such a scheme of procuring philosophical instruments, and systematising the lending out, and keeping in efficient order theodolites, levels, telescopes, etc.; and we concurred in opinion that the scheme could not succeed. We deemed it advisable rather to endeavour to get Short's observatory on the Calton Hill occupied as a 'Popular Observatory.' We spoke to some of the magistrates on this subject, who, on the part of the town, were quite favourable to the idea. We also applied to Mr. Thomas Allan, then an active member of the Royal Society, and he joined us in a communication to Sir

George Mackenzie of Coul, who warmly entered into our views; and ultimately we had an interview with Professor Playfair, who, in his mild and placid manner, agreed to consider the subject, but felt some difficulty on account of his colleague, the Professor of Practical Astronomy. After a time Professor Playfair undertook to draw up a statement for the public, which he did in his usual elegant and concise style. Thus, step by step, we succeeded in obtaining subscribers, and under the countenance and support of Playfair, many were found who patronised the proposal of establishing an observatory on the Calton Hill.

"Our idea was that we might look forward to a Popular Observatory which would not interfere with the existing Professorship of Astronomy, but have an establishment to which, with our families, we might resort in an evening with the advantage of oral and ocular demonstrations in the science of Astronomy, treated after a popular form.

"The present characteristic and beautiful building was then erected, and with the aid of Government, it was furnished with some of the chief instruments; but much to my regret the establishment has been exclusively limited to the purposes of a scientific observatory, without any provision of a popular description for which it was originally intended.

"Unfortunately there was nothing to keep our constitution alive in the minds of the public—nothing to allure additional subscribers to our funds, so as to extend the building, and fit it with a theatre and apparatus for popular purposes—no Lecture was established, and, in short, the original object fell dead in the hands of the

Directors. I thus personally lost my object in this establishment, and in all my *uphill* journeys and manifold
meetings, I had chiefly in view the pleasure of interviews
with my excellent friend the late Thomas Henderson, the
Professor of Astronomy in the University."

Passing from what may be regarded as Mr. Stevenson's public character as an engineer, it is only natural
that I should conclude this Memoir by adding a few
paragraphs descriptive of his social bearing as a man.

In politics my father was a decided conservative, but
he never took a prominent part in political or municipal
affairs. He was, however, from his earliest days a loyal
subject of the king; and, as we find from his Journal, a
zealous supporter of the Government. He says:—"After
my return from the Pentland Skerries in 1794, I enrolled
myself as a private in the 1st Regiment of Edinburgh
Volunteers raised as the local Defenders of our *Firesides*
against the threatened invasion by the French, and served
about five years in the ranks of that corps. However,
when the war became hot, and invasion was fully expected,
other corps of Volunteers were embodied, when I was
promoted to be a Lieutenant in the ' Princess (Charlotte's)
Royals,' and afterwards Captain of the Grenadier Company."

His connection with the volunteers seems to have
been of a very agreeable and satisfactory character,
proving that such loyal and patriotic services were not
then and are not now incompatible with the most ardent
pursuit of those studies and duties which are to qualify a

man for the business of life. On his promotion to the
Royals he received the following friendly letter from his
Colonel, Charles Hope, Lord Advocate, and afterwards
Lord President of the Court of Session :—

"*24th January* 1804.

"Sir,—I always part with any of my friends in the Regiment
with great regret, especially such as belonged to the old Blues.
But I cannot object to your leaving me in order to be more extensively useful in another corps. I therefore heartily wish you every
success in your new undertaking, and have no doubt that you will
prove a valuable acquisition to the discipline of the Spearmen.

"Notify to Captain Spens your resignation, that he may send
for your arms.—I am, Sir, yours sincerely,

C. Hope,
"*Lt.-Col. 1st. R.E.V.*

"Mr. Robert Stevenson,
"Capt., Spens' Company."

Mr. Stevenson remained several years in his new
corps, until he was obliged, on commencing the Bell Rock
Lighthouse, to tender his resignation, when he received
a letter from Colonel Inglis conveying the request
of the Regiment that he should continue as an honorary
member of the corps :—

"Edinburgh, *9th April* 1807.

"Sir,—My anxious desire to have, if possible, devised means
for detaining you among us, must plead my excuse for being so
long of replying to your letter ; and it is with most sincere regret,
that, after the most mature consideration, I am obliged to express
my fears that the rules of the Volunteer Corps must deprive us of

your services, in consequence of your active charge of a work of national importance, rendering your absence from Edinburgh unavoidable for years, during the months of drill.

"While I feel myself impelled, therefore, to accept of your proffered resignation, I beg to assure you of my own sense, as well as that of all the other officers, of the loss we sustain, and of our great personal regard.

"And I am directed to entreat you will do us the favour of continuing as an honorary member of a corps which has been so much indebted for your zeal and exertions.

"I cannot conclude without returning you my thanks for the obliging sentiment contained in your letter towards myself; and have the honour to be, with much esteem, sir, your faithful obedient servant,

WILLIAM INGLIS, L.C.C., L.E.S.

"CAPTAIN STEVENSON, ETC."

Many of his personal friends have recorded the pleasant satisfaction with which they continued through life to look back upon the days spent in my father's company on board the lighthouse tender, while making his annual inspection of the lighthouses. On one of these voyages he was accompanied by his friends Patrick Neill, LL.D., the Botanist; Charles Oliphant, Writer to the Signet; and John Barclay, M.D., the Anatomist; who presented him with a piece of plate in remembrance of "the many happy hours they passed in his company on sea and shore."

On another occasion in 1814, the Commissioners of Northern Lighthouses invited Sir Walter Scott to accompany them on their annual tour. Mr. Lockhart, in his life of Scott, says, "The company were all familiar friends of his, William Erskine, then Sheriff of Orkney, Robert Hamilton, Sheriff of Lanarkshire, Adam Duff, Sheriff of Forfarshire, but the real chief of the expedition was the Surveyor Viceroy, the celebrated Engineer Stevenson, and Scott anticipated special pleasure in his society." "I delight," Scott writes to Morritt, "in these professional men of talent; they always give you some new lights by the peculiarity of their habits and studies, so different from the people who are rounded, and smoothed, and ground down for conversation, and who can say all that every other person says, and—nothing more." I quote a single paragraph from Scott's diary of this memorable voyage, in which he gives an amusing account of the first landing of the Commissioners on the rock on which the celebrated Skerryvore lighthouse has since been erected by Alan Stevenson, who succeeded my father as Engineer, on his retirement from the Scottish Lighthouse Board in 1843.

" Having crept upon deck about four in the morning," says Sir Walter, " I find we are beating to windward off the Isle of Tyree, with the determination, on the part of Mr. Stevenson, that his constituents should visit a reef of rocks called Skerry Vhor, where he thought it would be essential to have a lighthouse. Loud remonstrances on the part of the Commissioners, who one and all declare they will subscribe to his opinion, whatever it may be, rather than continue this infernal buffeting. Quiet perseverance on the part of Mr. S., and great kicking, bouncing, and squabbling upon that of the yacht, who seems to like the idea of Skerry Vhor as little as the Commissioners. At length, by dint of exertion, come in sight of this long ridge of rocks (chiefly under water), on

which the tide breaks in a most tremendous style. There appear
a few low broad rocks at one end of the reef, which is about a mile
in length.   These are never entirely under water, though the surf
dashes over them.   To go through all the forms, Hamilton, Duff,
and I resolve to land upon these bare rocks in company with Mr.
Stevenson.   Pull through a very heavy swell with great difficulty,
and approach a tremendous surf dashing over black, pointed rocks.
Our rowers, however, get the boat into a quiet creek between two
rocks, where we contrive to land well wetted.   I saw nothing
remarkable in my way excepting several seals, which we might
have shot, but in the doubtful circumstances of the landing, we
did not care to bring guns.   We took possession of the rock in name
of the Commissioners, and generously bestowed our own great
names on its crags and creeks.   The rock was carefully measured
by Mr. S.   It will be a most desolate position for a lighthouse,
the Bell Rock and Eddystone a joke to it, for the nearest land is
the wild island of Tyree, at fourteen miles' distance.   So much for
the Skerry Vhor."

In family life Mr. Stevenson was a man of sterling
worth.   As a husband, a father, and a friend, he was
remarkably distinguished by the absence of selfishness.
His exertions in forwarding the progress of young men
through life were generous and unwearied; and few
men had more solid grounds than he for indulging in the
pleasing reflection that, both in his public and private
capacity, he had consecrated to beneficial ends every talent
committed to his trust.

He was a man of sincere and unobtrusive piety;
and although warmly attached to the Established Church
of Scotland, of which for nearly forty years he had been
an elder, and for many years a member of the General

Assembly, he had no taint of bigotry or of party feeling, and he died calmly in that blessed hope and peace which only an indwelling personal belief in the merits of a Redeemer can impart to any son of our race.

At a statutory general meeting of the Board of Northern Lighthouses, which was held on the 13th July 1850, the day after my father's death, the Commissioners recorded their respect for his talents and virtues in the following Minute :—

"The Secretary having intimated that Mr. Robert Stevenson, the late Engineer to the Board, died yesterday morning,

"The Board, before proceeding to business, desire to record their regret at the death of this zealous, faithful, and able officer, to whom is due the honour of conceiving and executing the great work of the Bell Rock Lighthouse, whose services were gratefully acknowledged on his retirement from active duty, and will be long remembered by the Board, and to express their sympathy with his family on the loss of one who was most estimable and exemplary in all the relations of social and domestic life. The Board direct that a copy of this resolution be transmitted to Mr. Stevenson's family, and communicated to each Commissioner, to the different lightkeepers and the other officers of the Board."

APPENDIX.

THE INCHCAPE ROCK.

AN old writer mentions a curious tradition, which may be worth quoting. " By east the Isle of May," says he, " twelve miles from all land in the German Seas, lyes a great hidden rock, called Inchcape, very dangerous for navigators, because it is overflowed everie tide. It is reported in old times, upon the saide rock there was a bell, fixed upon a tree or timber, which rang continually, being moved by the sea, giving notice to the saylers of the danger. This bell or clocke was put there and maintained by the Abbot of Aberbrothok, and being taken down by a sea pirate, a yeare therafter he perished upon the same rocke, with ship and goodes, in the righteous judgement of God."—STODDART'S *Remarks on Scotland*.

No stir in the air, no stir in the sea,
The ship was still as she could be;
Her sails from heaven received no motion,
Her keel·was steady in the ocean.

Without either sign or sound of their shock
The waves flowed over the Inchcape Rock;
So little they rose, so little they fell,
They did not move the Inchcape Bell.

The Abbot of Aberbrothok
Had placed that Bell on the Inchcape Rock;
On a buoy in the storm it floated and swung,
And over the waves its warning rung.

When the Rock was hid by the surge's swell,
The mariners heard the warning Bell;
And then they knew the perilous Rock,
And blest the Abbot of Aberbrothok.

The Sun in heaven was shining gay,
All things were joyful on that day;
The sea-birds scream'd as they wheel'd round,
And there was joyaunce in their sound.

The buoy of the Inchcape Bell was seen
A darker speck on the ocean green;
Sir Ralph the Rover walk'd his deck,
And he fix'd his eye on the darker speck.

He felt the cheering power of spring,
It made him whistle, it made him sing;
His heart was mirthful to excess,
But the Rover's mirth was wickedness.

His eye was on the Inchcape Float;
Quoth he, "My men, put out the boat,
And row me to the Inchcape Rock,
And I'll plague the Abbot of Aberbrothok."

The boat is lower'd, the boatmen row,
And to the Inchcape Rock they go;
Sir Ralph bent over from the boat,
And he cut the Bell from the Inchcape float.

Down sunk the Bell with a gurgling sound,
The bubbles rose and burst around;
Quoth Sir Ralph, "The next who comes to the Rock
Won't bless the Abbot of Aberbrothok."

Sir Ralph the Rover sail'd away,
He scour'd the seas for many a day;
And now grown rich with plunder'd store,
He steers his course for Scotland's shore.

So thick a haze o'erspreads the sky
They cannot see the Sun on high;
The wind hath blown a gale all day,
At evening it hath died away.

On the deck the Rover takes his stand,
So dark it is they see no land.
Quoth Sir Ralph, "It will be lighter soon,
For there is the dawn of the rising Moon."

"Canst hear," said one, "the breakers roar?
For methinks we should be near the shore."
"Now, where we are I cannot tell,
But I wish I could hear the Inchcape Bell."

They hear no sound, the swell is strong;
Though the wind hath fallen they drift along,
Till the vessel strikes with a shivering shock,
"O Christ! it is the Inchcape Rock!"

Sir Ralph the Rover tore his hair;
He curst himself in his despair;
The waves rush in on every side,
The ship is sinking beneath the tide.

But even in his dying fear
One dreadful sound could the Rover hear,
A sound as if with the Inchcape Bell,
The Devil below was ringing his knell.

INDEX.

PRINTED BY T. AND A. CONSTABLE, PRINTERS TO HER MAJESTY,
AT THE EDINBURGH UNIVERSITY PRESS.

Printed in the United States
By Bookmasters